T0258220

MUCH LIKE US

What Science Reveals about the Thoughts, Feelings, and Behaviour of Animals

What really differentiates us from our relatives in the animal world? And what can they teach us about ourselves? Taking these questions as his starting point, Norbert Sachser presents fascinating insights into the inner lives of animals, revealing what we now know about their thoughts, feelings, and behaviour. By turns surprising, humorous, and thought-provoking, *Much Like Us* invites us on a journey around the animal kingdom, explaining along the way how dogs demonstrate empathy, why chimpanzees wage war, and how crows and ravens craft tools to catch food.

Sachser brings the science to life with examples and anecdotes drawn from his own research, illuminating the vast strides in understanding that have been made over the last 30 years. He ultimately invites us to challenge our own preconceptions – the closer we look, the more we see the humanity in our fellow creatures.

Norbert Sachser is Professor of Zoology and Head of the Department of Behavioural Biology at the University of Münster, Germany. Widely considered to be one of Germany's leading behavioural biologists, he is known particularly for his pioneering work on stress, social behaviour, and welfare indicators in mammals.

MUCH LIKE US

*What Science Reveals about the Thoughts, Feelings,
and Behaviour of Animals*

Norbert Sachser

University of Münster

Translated from the German by Ruby Bilger

CAMBRIDGE
UNIVERSITY PRESS

CAMBRIDGE
UNIVERSITY PRESS

University Printing House, Cambridge CB2 8BS, United Kingdom

One Liberty Plaza, 20th Floor, New York, NY 10006, USA

477 Williamstown Road, Port Melbourne, VIC 3207, Australia

314–321, 3rd Floor, Plot 3, Splendor Forum, Jasola District Centre,
New Delhi – 110025, India

103 Penang Road, #05–06/07, Visioncrest Commercial, Singapore 238467

Cambridge University Press is part of the University of Cambridge.

It furthers the University's mission by disseminating knowledge in the pursuit of
education, learning, and research at the highest international levels of excellence.

www.cambridge.org
Information on this title: www.cambridge.org/9781108838498
DOI: 10.1017/9781108975001

Original Title: Der Mensch im Tier

Copyright © 2018 by Rowohlt Verlag GmbH, Reinbek bei Hamburg

English edition translated by Ruby Bilger

© Cambridge University Press 2022

First published 2022

Printed in the United Kingdom by TJ Books Limited, Padstow Cornwall

A catalogue record for this publication is available from the British Library.

ISBN 978-1-108-83849-8 Hardback

For Claudi

Tat tvam asi

These words in Sanskrit were painted on the wall of the animal housing room in the famous evolutionary biologist Bernhard Rensch's institute at the University of Münster more than 50 years ago, according to his student Gerti Dücker. They mean: 'This is you.'

Contents

Preface

Most of us are interested in animals from a young age. Their behaviour fascinates us. The frequency with which their activities are covered in print media, television, and the web only reinforces this interest: humans want to pay attention to what animals are doing around us. Indeed, what society thinks of animals – how we interact with them, and how we interpret and explain their behaviour – changes with time. And in the last few years we have experienced a fundamental transformation.

The discipline largely responsible for this change is behavioural biology, which describes animal behaviour and seeks to identify its underlying causes and consequences. This book is for all who are interested in animal behaviour and the evolution of the scientific concept of the animal, and who would like to understand what current science actually knows about their thoughts, emotions, and behaviour.

It was a long road to finishing this book. The ideas at its core first emerged in the mid-1990s when Rainer Hagencord invited me to give a lecture to the Catholic community of the University of Münster. Conscious of the increasing ecological and bioethical problems facing science and our society at large, he had made it his mission to advance the interdisciplinary dialogue between the natural sciences and theology and philosophy. I chose the theme, 'Humans: The Pride of Creation? On the Thoughts, Feelings, and Behaviour of Animals'. It was here that I developed, through data and arguments from behavioural biology, the central idea of this book: humans have grown closer to animals; there is much more of us in animals than we could have even a few years ago imagined. At that time, I had no idea how much this thesis would be supported by new findings in behavioural biology in the coming years.

The original German title of this book – *Der Mensch im Tier* (literally, *The Human in the Animal*) – goes back to a project of the same name at the UniKunstTage 2000 in Münster, which was initiated by my colleague Reinhard Hoeps, and sought to bring art into conversation with the natural sciences. The interaction between the artists and us biologists not only led to remarkable works – Silke Rehberg's 'Guinea Pig in Blue', a set of circular reliefs of glazed terra-cotta, have since been prominently displayed on the front of our institutional building – but also raised my awareness to the fact that there is not only much of animal nature to be found in humanity, as is often said, but also much of human nature to be found in animals. Since then, I have found the converse perspective much more exciting.

The publication of this book, so many years later, is thanks to the persuasiveness of my editor, Frank Strickstrock. He became aware of a statement I had originally made in a conversation with *Der Spiegel*: 'We are currently experiencing a revolution of the concept of the animal,' and, during a visit to Münster, he asked if I could imagine writing a book on this topic. Although I was hesitant at first, after several follow-up meetings, my excitement grew.

And here is that book! I cover six topics from behavioural biology that are central to the change in the scientific understanding of the animal and have helped shrink the conceptual gap between humans and animals. To be clear, my personal research interests are also included in the selection of topics and can, of course, only reflect a small portion of current research in the discipline. Readers should note that each chapter of this book can stand alone. Those who are most interested in the topic of emotions and animal welfare should start with Chapter 3, while those who prefer to focus on animal personalities can begin with Chapter 6.

I could not have found this path in science alone, and this book would not exist without the support of others, so I have many people to thank! My parents nurtured my interest in research from a young age and have supported me unconditionally along the way. My teachers and mentors, first and foremost Klaus Immelmann, Hubert Hendrichs, and Dietrich von Holst, shaped me through their examples and showed me what 'good science' is. Our research in the past few years would not have been possible without the excellent members of my team, many of whom are now professors themselves or hold other important positions. The scientific exchange with other researchers from all over the world has been indispensable. Many thanks to my colleagues at the Münster Graduate School of Evolution with whom I have had many stimulating discussions throughout the years, often far beyond the boundaries of my own subject.

I would also like to thank the German Research Foundation for having generously funded our research for decades. Our results, which are described in Chapter 2, were funded by the 'Social Physiology' project, and a number of the studies described in Chapters 3 and 4 are based on our project, 'Fear, Anxiety, and Anxiety Disorders', which we carried out through the Collaborative Research Centre. We gained many of the insights in Chapter 6 as part of the research group, 'Early Experience and Behavioural Plasticity', and the special research programme, 'The Individual and Its Ecological Niche', and our work described in Chapter 7 was carried out in the Priority Programme, 'Genetic Analysis of Social Systems'.

When the first version of this book was finished, a number of esteemed behavioural biologists agreed to give each chapter a careful read through. Many thanks to Oliver Adrian, Rebecca Heiming, Niklas Kästner, Sylvia Kaiser, Helene Richter, and Tobias Zimmermann. I would also like to thank my wife, Claudia Böger. As a doctor of humanities, she has been a part of my research in an interdisciplinary and constructive manner for more than three decades. Her critical reading of the manuscript and her many helpful suggestions were an essential contribution to the creation of this book.

CHAPTER 1

Typical Human, Typical Animal?

Reconceptualising the Animal – an Introduction

A REVOLUTION HAS RECENTLY TAKEN PLACE IN BEHAVIOURAL biology. Its consequences are far-reaching, both for our self-image as humans and for our relationship with animals. Just a few decades ago, behavioural science was guided by two key dogmas: animals cannot think, and no scientific statements can be made about their emotions. Today, the same discipline holds both ideas to be false and posits the very opposite: animals of some species are capable of insight – they can recognise themselves in a mirror and exhibit at least a basic sense of self-awareness – and they have rich emotional lives that seem to be startlingly similar to those of humans. Situations that lead to strong emotional responses in humans, whether positive or negative – for example, when we fall in love or lose a partner – seem to have the same effect on our animal relatives.

Indeed, the transformation of the concept of the animal in modern behavioural biology has been so fundamental that it amounts to a paradigm shift. And since it has long since become untenable to distinguish between *Homo sapiens* as driven by reason and animals as driven by instinct, the question arises: what actually differentiates humans from animals? How much of ourselves is present in them?

The general perception of these differences has also changed in parallel to developments in the life sciences. A few decades ago, if biology students had been presented with photos of a goldfish, a chimpanzee, and a human, and asked to sort them into two categories of their own devising, more than 90 per cent would have put the human in the first category and the 'animals' in the second. If biology students today are asked the same question in their first semester, the result is completely different: over 50 per cent group humans and chimpanzees into one category and the goldfish into another. Apparently, humans and animals have grown closer to one another in the public imagination, too.

This has been confirmed by the death of a third dogma: for decades, it was taught that animals behave for the good of their species, generally never killing members of their own – known as 'conspecifics' – and often helping them to the point of self-sacrifice. Today we know that this is not the case. Rather, animals do everything to ensure that copies of their own genes are passed to the next generation with maximum efficiency and, when necessary, they will also kill conspecifics. Clearly, they are not, as Jane Goodall had once famously hoped, 'like us, but better'.

The border between humans and animals is also beginning to blur in other areas. Certain aspects of the social environment can cause stress for both humans and animals, while other similar factors can alleviate it. Both have their thinking, feelings, and behaviour shaped by similar interactions between genetics and environment. Indeed, animal behaviour does not develop in a fixed manner: environmental influences, socialisation, and learning can alter an animal from the prenatal phase through adulthood. Like humans, animals ultimately appear individualised upon closer inspection, which is why behavioural biology now takes animal personalities into account.

This book will demonstrate how and why the scientific understanding of animal behaviour has changed so fundamentally. It will focus on a group of animals to which, biologically speaking, we also belong: mammals, whose approximately 5500 species populate the most diverse range of habitats on our planet. Lions and zebras inhabit the savannah, gorillas and orangutans inhabit the tropical rainforests, fennecs live in deserts, polar bears live in the arctic, moles live underground, bats have taken to the skies, and whales and seals have taken to the water.

Humans have much in common with all these creatures. For one, our genes: we share about 99 per cent of them with our closest relatives, bonobos and chimpanzees. Brain structure is also nearly identical across all mammals: the so-called 'ancient' parts of the brain in particular – like the limbic system – show similarities down to the last detail. A human's fear response at the sight of a snake, for instance, may likely be controlled by the exact same neuronal process as in a chimpanzee or a squirrel monkey. Our physiological regulatory systems, too, are strikingly similar. The same hormones enable all mammals to cope with stressful situations, adapt to changing environments, or reproduce. In fact, the production of the sex hormones testosterone and oestradiol, the stress hormones adrenaline and cortisol, or the 'love' hormone oxytocin is not unique to humans; rather, these hormones occur in the same form in a wide variety of species, from bats to rhinoceroses to dolphins.

However, such similarities across genes, brain structure, and the endocrine system do not automatically imply similarities concerning thoughts, feelings, and behaviour. To better understand these traits, we need to look at specific studies in both animals and humans. In the case of animals, such studies take place within the field of behavioural biology, which was aptly defined by one of the fathers of the discipline, the Nobel Prize winner Nikolaas Tinbergen, as 'the study of behaviour by biological methods'.

THE STUDY OF BEHAVIOUR BY BIOLOGICAL METHODS

This definition can very simply be illustrated by the relationship between a general knowledge of animals and a knowledge of behavioural biology: a knowledge of animals is certainly required to study behavioural biology, but it is not in itself sufficient to draw scientific conclusions about animal behaviour. Thus, the terms are by no means synonymous. Not everyone who interacts with animals and makes statements about their behaviour is a behavioural biologist, although people who have close contact with animals may have an excellent knowledge of their behaviour. My grand-mother, for example, was always right about our dog – if she warned he was about to bite, one did well to take it seriously. But this was not knowledge in a scientific sense: it was intuition acquired through experi-ence, and, if I had asked her how she knew these things, she would have answered, 'I can just tell.' Experiential or intuitive knowledge can, of course, be just as true as scientific knowledge. But it does not have to be, so it is very hard to decide when it is valid and when it is not. Take, for example, the characteristics of certain animals that have made their way into the vernacular: we speak of the thieving magpie or the silly goose; we compare a clumsy person to a bull in a china shop, or a person stuck in their habits to an old dog who cannot learn new tricks. Whether these attributes are accurate to these animals or not can ultimately only be clarified through behavioural studies, which have indeed frequently shown them to be prejudices.

How, then, is knowledge characterised in behavioural biology? As with any type of scientific knowledge, it must be possible to convey the methods and procedures by which it was acquired. This was not the case with my grandmother's knowledge of our dog – it is not enough for someone to sit in front of a group of animals, be affected by their behaviour, and describe his or her subjective impressions of it. In a legitimate behavioural study, the researcher must first list and define the specific behaviours of the animal species under study in what is known as an ethogram. Then, data is collected on these behaviours using an appropriate method: if the researcher were studying animal social life, for instance, he or she would record how often and for how long each animal exhibited socio-positive behaviour (that is, friendly behaviour towards other members of the group), how often each animal initiated or was the target of aggressive behaviour, how often each animal positioned itself next to certain others in the group, and which males paired with which females. These observations used to be collected by hand, but behavioural data is now recorded and analysed with sophisti- cated software, as is the statistical evaluation of the results.

Let's stay with the topic of mammalian social life for a little while longer. The history of research in this area also shows how crucial it is to use the right method of data collection. A few decades ago, when the first of such studies were being conducted in animals' natural habitats, scien- tists often used the *ad libitum* method: they observed all the animals in a group simultaneously and recorded all behaviours that they noticed. However, this method introduced a huge problem that has long been known to perceptual psychology: humans tend to focus their attention on what is loud and distinctive, neglecting events that occur quietly and unobtrusively. In many mammalian societies, male behaviour – especially in interactions with conspecifics – is more expressive and louder than female behaviour, as confrontations with other males are often marked by conspicuous vocalisation. If researchers apply the *ad libitum* method to these interactions, they will inevitably collect significantly more data on males than females. This perceptual bias may have contributed to the fact that males have long been described as dominant and tone-setting in many mammalian societies, while females have been characterised as passive and inferior.

After this procedural error was recognised, researchers began to replace the *ad libitum* method with what is known as focal-animal sampling, in which each animal in the group is observed for the same amount of time regardless of what it is doing, thus ensuring that all are given the same amount of attention. The data collected using this method has contributed significantly to the revision of our concept of the female role in mammalian societies: we know today that females are by no means passive, but rather tend to interact in subtler yet no-less-influential ways. Recent behavioural biology textbooks reflect this insight, teaching that it is often the females in primate societies who make the most important decisions for the group.

While animal societies like those of primates are organised into fixed groups of several adult males and females, there is a great diversity in the social life of mammals: many species, like tigers, live solitary lives, while others, like certain zebras, organise themselves into harems, and elephants, who constitute the strongest matriarchy in the animal kingdom, present close, sometimes lifelong, bonds between the females of a group. Such long-term bonds between males are found in a few species, such as the cheetah. In the saddle-back tamarin, a small South American species of monkey, harems of one female and two males regularly occur. Interestingly, the favoured lifestyle of humans – monogamy – rarely occurs in non-human mammals: no more than 3–5 per cent of species organise themselves into pairs. (One that does is the North American prairie vole.) None of our closest biological relatives – bonobos, chimpanzees, gorillas, or orangutans – live monogamous lives.

Given this great variety of species, habitats, and lifestyles, studies in behavioural biology must not only be conducted using a sound methodology but the results must also be reproducible. If a research group in

Berlin shows that bees can orient themselves to the position of the sun, for example, then this result must also be obtainable by researchers in London or Tokyo.

The importance of reproducibility is wonderfully illustrated by a certain well-known historical case study. Shortly before the First World War, a man named Wilhelm von Osten caused quite a stir with his horse, Clever Hans. Hans could seemingly do basic arithmetic – addition, subtraction, and division – and indicate the correct answers to problems by stomping on the ground or nodding his head. The public was amazed, but scientists quickly began to doubt that a horse was capable of such a mental feat. Wilhelm von Osten agreed to an investigation, and indeed, the first study showed that Clever Hans was able to solve basic equations even if they were given to him by strangers. However, as the study continued, it was revealed that Clever Hans could no longer solve a problem if no one present knew the solution. The horse, it turned out, was able to pick up on the smallest nuances in the body tension of the person who gave him the mathematical problem to deduce when he should stop stomping or nodding. Clever Hans had extraordinary sensory perception – but he could not do arithmetic.

Nevertheless, he has had a lasting impact on research. Today it is generally accepted that displays of animal cognition can only be scientifically verified through so-called blind studies, during which the experimenter does not know the solution to the task given to the animal. Unconscious assistance, which must be eliminated in any legitimate study, is known as the 'Clever Hans effect'. Wilhelm von Osten was certainly no charlatan – he was firmly convinced of his horse's cognitive abilities. Even today, many pet owners attribute outstanding cognitive abilities to their dogs or cats, claiming things like: 'My dog understands every word I say.' Whether this is really the case, however, cannot be judged from everyday experience alone. Clever Hans has certainly taught us that.

The basic behavioural biological method is therefore known as the process of objectively and reproducibly recording animal behaviour. Depending on the study, however, techniques from neighbouring disciplines may also be used. Researchers rely on state-of-the-art satellite technology to determine the position of birds during migration, for example; they divine the stress state of animals by measuring their hormone levels; they determine paternity or kinship with the help of molecular genetics. Such techniques allow scientists to gain insights that would not be possible through behavioural observation alone, which can often be misleading. For example, songbirds have long been considered in the public imagination as the epitome of fidelity. But paternity verification through genetic fingerprinting has revealed a very different picture: a large part of the offspring found in these birds' nests often do not come from the males who occupy them and feed the young there. Evidently, songbirds are not as monogamous as humans might like to believe.

A SHORT HISTORY OF BEHAVIOURAL BIOLOGY

Since their earliest days of existence, humans have taken an interest in the animals that surround them: in how to escape them, hunt them, or even just enjoy their presence. The cave paintings at Altamira and Lascaux, which are among the oldest works of art in human history, are a Stone Age testimony to the human–animal relationships of this early period. Through breeding, we have been domesticating what were once wild animals for thousands of years: sheep, pigs, cattle, and goats have lived among us for about 10 000 years, while dogs may have been man's faithful companions for as long as 30 000 years.

Greek philosophers began to contemplate the nature of humans vis-à-vis animals about 2500 years ago. Aristotle famously saw the animal's lack of reason as a fundamental difference between the two, and this distinction is still anchored in much of society's consciousness today: many still believe that humans alone possess reason while animals can only follow their instincts.

The first examples of empirical scientific and experienced-based observation of animal behaviour can be found in the Middle Ages. In the thirteenth century, Emperor Frederick II, known to his contemporaries as *stupor mundi*, 'the wonder of the world', wrote *De Arte Venandi cum Avibus/ The Art of Hunting with Birds*, which can be considered the first scientific book of western ornithology – or, some may argue, of behavioural biology. As early as the sixteenth century, naturalists such as Konrad Gesner, Carl von Linné, and Jean-Baptiste de Lamarck were describing and systematising animals and plants, including many species from the parts of the world only recently visited by Europeans. These writings include numerous descriptions and contemplations of animal behaviour, but general consensus does not consider behavioural biology to have truly emerged as a discipline until the middle of the nineteenth century.

The father of behavioural biology (and many other related disciplines) is the British naturalist Charles Darwin. In *On the Origin of Species*, first published in 1859, Darwin lays out the basic features of his theory of evolution, which we still hold to be true today. He understood evolution to be two things: first, the process by which species change over time, based on the premise that they do not exist in a static state but rather are altering their appearance and behaviour constantly. The second feature is the descent from common ancestors. Eight to ten million years ago, for example, there were no humans or chimpanzees on our planet. There did exist, however, a certain species of ape, now extinct, from which both humans and the chimpanzee derive. Through his studies, Darwin not only proved that evolution exists, but also recognised the major driving force behind evolutionary change: natural selection.

What does this key concept mean? Darwin recognised that all organisms have a nearly unlimited ability to reproduce – many more offspring can be created in a single generation than there are parents. But this enormous potential is not realised; rather, the size of a population remains more or less constant, meaning that the majority of offspring perish. Only a few survive to sexual maturity, and even fewer subsequently reproduce. Therefore, Darwin posits, there must be steep competition for survival and scarce resources such as food, habitats, and mates: what he termed the struggle for existence. Which animals survive is by no means left to chance. Individuals who are better adapted to their environment through hereditary advantages – for example, they find food or mates more easily or are more likely to escape predators – are more likely to survive and successfully reproduce than their less-capable conspecifics. The genetic makeup that allowed certain individuals to survive is then successfully passed on to the offspring, while the genetic makeup that caused others to perish is lost. Through this process of natural selection, animal species become constantly better adapted to their environment.

One chapter of *On the Origin of Species* is devoted exclusively to animal behaviour. In it, Darwin states that instincts and the behaviours they control, just like all other characteristics of an organism, are modified through natural selection and therefore continually adapted to the environment. He thus anticipates a central theme of behavioural ecology, an important discipline of contemporary behavioural research: the adaptation of behaviour to ecological conditions. He further describes the similarities that appear between the instincts of closely related species that are present even when they live in separate parts of the world. Both South American and European species of thrushes, for example, line their nests with mud. That closely related species share more common behaviours in their ethogram than distantly related ones would become a central dogma of comparative behavioural research decades later.

In 1872, Darwin published another book: *The Expression of the Emotions in Man and Animals*. In it, he argues that certain facial expressions – especially those that reflect basal emotions such as joy, sadness, or anger – exist independently of culture and are thus innate. Furthermore, he says, some animal species may possess emotions comparable to those of humans, which they express using similar faces. The book became a bestseller shortly after it was published, although it did not catch on in the scientific community and, for a long time after, was virtually forgotten. Then, in the 1960s, the biologist Irenäus Eibl-Eibesfeldt revisited Darwin's theses and founded human ethology, a subdiscipline of behavioural biology that attempts to comprehend emotions as innate features of human behaviour. Indeed, Eibl-Eibesfeldt was able to identify universal similarities in human facial expressions when he compared feelings such as joy, sadness, or disgust among different ethnic groups across Africa, South America, and Asia.

At the time, animal emotions had not been a topic in behavioural biology for well over a century – the idea that humans and animals shared certain emotions had long been considered politically incorrect. But in the last decade or so this has changed dramatically. Today, emotions are a central field of research in behavioural biology, and perhaps in this context we will see a Renaissance of Darwin's long-forgotten work.

For about half a century after Darwin, the majority of biologists were not specifically interested in animal behaviour: research tended to focus on systematics, physiology, and developmental biology. Only then did the field we now call behavioural biology begin to emerge through the writings of the researchers Konrad Lorenz, Nikolaas Tinbergen, and Karl von Frisch.

Karl von Frisch studied sense perception: how animals calibrate themselves to their environment and communicate with one another. He was the first to prove that fish can hear and bees can see colour, and that they orientate themselves with the help of a solar compass. von Frisch became known primarily through his investigations into animal communication, in which he discovered the so-called 'waggle dance' used by individual bees to tell hive-mates the direction and distance of a food source. von Frisch was also the first scientist to study animal behaviour through a logical sequence of related experiments.

While von Frisch is a key figure in behavioural biology (or ethology or animal psychology, as it was also called in its early days), the emergence of the discipline was even more influenced by the researchers Konrad Lorenz and Nikolaas Tinbergen. Through their work, it was first accepted that behaviour can be studied in the same way that anatomy, morphology, or physiology can, and observation of animal behaviour was established as a serious scientific method. In a series of classic studies, Lorenz described the behaviour of various duck species down to the smallest possible units, which were termed 'fixed action patterns'. These relatively stereotyped behaviour patterns are exhibited by all members of the same species, at least those of the same age and sex: one could say that the courtship behaviour of a mallard in Berlin is the same as that of one in Beijing. Lorenz's comparison of fixed action patterns across different species such as mallards, Meller's ducks, pintails, shovelers, teals, wigeons, or mandarin ducks in turn showed that the more closely certain species were related, the more fixed action patterns they shared. Thus, comparative ethology was born.

Through observing ducks and geese, Lorenz also recognised that these animals have no innate knowledge of their species' appearance – rather, they only learn to recognise each other through what is known as imprinting. In a specific window of time shortly after hatching, chicks will become fixated on whatever moves and makes noise in their vicinity. In their natural habitat this is usually the mother, whom the chicks then learn to follow. But if, during this phase, Lorenz moved the chicks around and called to them instead, the chicks would irrevocably imprint on him. If they later had a choice to follow him or their mother, the chicks would choose Lorenz.

Lorenz also developed important models for behavioural control. According to these, key environmental stimuli activate releasing mechanisms, which results in an associated innate behavioural response. Tinbergen was then able to experimentally prove that these models held true for many animal species. For example, if a rival invades a stickleback's territory, the stickleback will instinctually react with threatening behaviour. What causes this aggression? The red underside of the intruder's belly is the key stimulus: a lifelike stickleback model without a red underbelly does not trigger any aggression in the creature, but the stickleback will begin to violently attack a piece of wood whose lower half has been painted red, even though the wood does not remotely resemble a conspecific rival.

Tinbergen also conducted simple but ingenious experiments in the natural habitats of certain animals to understand the function of their behaviour. In one classic experiment, for example, he asked himself: Why do black-headed gull parents remove the broken eggshells from their nests after a chick hatches? To answer this question, he fashioned artificial nests with gull eggs, laying broken eggshells near some of the nests and leaving others without. After a while he noticed that the presence of an eggshell increased predation of a nest. Thus, parents apparently remove the eggshells to avoid predation. By studying the adaptive value of behaviour, Tinbergen laid the foundations for behavioural ecology, an important discipline of behavioural biology that emerged in the 1970s.

In these early days of behavioural biology, Lorenz, Tinbergen, von Frisch, and their ever-growing number of students studied many different species – birds, fish, and insects especially. The scientists were particularly fascinated by the fact that these animals seem to have an innate knowledge of how to behave and were thus perfectly adapted to their habitat. A digger wasp, for instance, knows how to seek the right kind of prey and build the right kind of nest without ever having had contact with its parents or learned from any conspecific. Like Darwin, the early behavioural biologists named the source of this knowledge 'instinct', and, like Darwin, they assumed that such instincts formed in the course of evolution through natural selection.

The importance of this topic is clear from the title of the first-ever textbook in behavioural biology: Tinbergen's *The Study of Instinct*, which first appeared in 1951. Indeed, the central goal of this early phase of behavioural biology was to study instinctive (that is, innate) behaviour. Several of the models developed at that time are no longer considered correct, such as the hierarchy of instincts, or the role of the interaction between environmental stimuli and internal factors in triggering instinctive behaviour. Nonetheless, the achievements of Lorenz, Tinbergen, and von Frisch cannot be understated: through their work, behavioural biology became an independent scientific discipline that has fundamentally changed the concept of animal behaviour. They were awarded the Nobel Prize in 1973.

Tinbergen in particular also pointed this fledgling field in the direction that it has taken up to the present day. In his article, 'On aims and methods of ethology', nearly 60 years ago, he provides the theoretical framework which is still the basis of behavioural biology: that explanations can and should be given for every behavioural phenomenon – from insect social organisation to chimpanzee tool use to bird song – on four different levels: mechanism, ontogeny, function, and phylogeny.

What does this mean? The reason why a male chaffinch sings, for example, can be explained in four different ways. The first is that the increasing length of the days in spring acts as an environmental stimulus that the male birds perceive, triggering the production of the sex hormone testosterone in their testes. This testosterone is then transported through the bloodstream to the brain, where it activates certain nerve impulses that direct the necessary muscles for singing. This is the causal explanation, which clarifies the mechanism of the behaviour.

A second explanation follows that a male chaffinch sings because he learned to from his father during a prescribed period when he was particularly able to do so. This is the life-historical explanation, which focuses on the ontogeny of the behaviour. (Ontogeny is understood as the period of time between the fertilisation of an egg and the death of an individual.)

A third explanation is that the male chaffinch sings to attract females and drive away potential rivals. This is the functional reason, which indicates what sort of adaptive value the behaviour has – why an animal who engages in this behaviour is better adapted to his or her environment and more successful in passing on his or her genes than a conspecific who does not.

Finally, the singing can be explained by the fact that a chaffinch is a songbird, descending from ancestors who sang. This is the phylogenetic explanation, which clarifies the behaviour based on its evolutionary history, or phylogeny. Tinbergen's point in laying out these four types of explanation was clear: we cannot understand a particular behaviour until we understand it on the level of its mechanism, function, ontogeny, and phylogeny, as well as the relationship of all four to each other. This message has never been more relevant to behavioural biology than it is today.

Indeed, in the past few decades, the questions Tinbergen posed about animal behaviour have been researched on a large number of species. But in the process behavioural biology has splintered into several disciplines that are unfortunately hardly related to each other: behavioural ecology and sociobiology, for instance, both focus on the functions of behaviour, its adaptive value, and its evolution, primarily asking: What are the advantages of certain behaviours? This line of research answers these questions wonderfully but neglects the ontogeny and mechanisms of the behaviour in doing so. These aspects are the focus of disciplines such as behavioural endocrinology, behavioural neurobiology, and behavioural genetics, which examine the relationship between behaviour and hormones, neurons, and genes, respectively. These disciplines mainly ask how a certain behaviour arises, but they are hardly concerned with its functions and evolution. Each of these disciplines has produced spectacular findings on animal behaviour, but there is very little unification of the results and the subsequent ways they have, taken together, changed our concept of the animal.

THE COMMON THREAD

So, here is where this book comes in: using these fundamental findings from different disciplines of behavioural biology, we can see how much the scientific concept of the animal has changed over the past decades.

This shift has undoubtedly been helped along by new emphasis on certain questions that were not addressed in the early days of the field: Can animals think? Do they have emotions? What about distinctive personalities? What does it really mean to be 'animal friendly?'

Furthermore, new methods have enabled us to re-examine old questions. For example, decoding the genomes of humans and certain animals has given us a much better understanding of the interplay between genes and environment in the execution of behaviour. Mammals, including many species of primates, have also increasingly become the subject of study, which has further contributed to our fundamentally altered view of animal behaviour. In summary, the findings of these various disciplines show that animals exhibit many characteristics, abilities, and behavioural patterns that we until recently regarded as unquestionably human.

In the following six chapters, this book presents the findings that have most notably contributed to this shift in understanding. The conclusion then summarises the new scientific concept of the animal and discusses how much of ourselves we might really see in it.

First, Chapter 2 deals with the relationship between behaviour and stress. The same characteristics of the social environment, it seems, lead to stress in humans and animals, while very similar factors can also reduce stress in both.

Chapter 3 addresses animal emotions and well-being, asking: What scientific methods can we use to determine whether animals are thriving? Under which conditions do they do well, and under which ones do they struggle? How do animals see the world? What do we know about their emotions? What exactly does an 'animal-friendly' life mean?

Chapter 4 looks at a question that has preoccupied science and society alike for many years: How much is behaviour determined by genes, and how much by environment? It traces how methods and perspectives on this problem have changed dramatically over the last few decades and shows how modern behavioural genetics provides new answers to old questions, culminating in the revolutionary realisation that genes not only influence behaviour but behaviour can also influence genes.

Chapter 5 deals with findings from cognitive biology, which address: How and what can animals learn? Can they think? Do some have self-awareness? And is it really true that great apes, our closest relatives, are more intelligent than other species?

Chapter 6 lays out mammalian behavioural development as an open process whose course is not predetermined at conception, birth, or even the end of childhood. A mammal's behaviour is already influenced by the environment in which its mother lives during the period of gestation, and the experiences the animal has throughout childhood and adolescence continually shape its behaviour. This is how animal personalities are formed, a subject of one of the newest areas of research in behavioural biology.

Chapter 7 addresses the central finding of sociobiology: animals apparently do not behave for the good of their species, but rather according to what are known as 'selfish genes'. If acting cooperatively helps them pass on their individual genetic material, then they will do it, but if their goal is better achieved through coercion or aggression – even to the point of killing conspecifics – then animals will exhibit such behaviour.

Chapter 8 concludes by reiterating the central point of this book: we, as humans, have moved much closer to animals than we have ever thought possible. Indeed, there is much more of us in them than we could have even a few years ago envisioned.

CHAPTER 2

Ginger Boris Doesn't Like to Be Alone

On Behaviour, Stress, and the Blessing of Socially Stable Relationships

HOW I CAME TO STUDY BEHAVIOURAL BIOLOGY

IN THE MID-1970S, I JOINED THE FIRST COHORT OF STUDENTS studying biology at the newly founded university in Bielefeld. In those days there were only about 30 to 40 of us students and about three professors, one of whom was Klaus Immelmann. Professor Immelmann had recently been called to serve as the first chair for animal behaviour at a German university, and his goal was to make Bielefeld a hub of behavioural biology research and teaching. He quickly succeeded. The facilities in his department were phenomenal, even by international standards. It was an exciting time: the campus was populated by various species of finches, parrots, geese, marmosets, kangaroos, deer, and rodents living in spacious indoor and outdoor enclosures. The founders of behavioural biology, Lorenz, Tinbergen, and von Frisch, had just received the Nobel Prize. There was a spirit of optimism in our eastern province of Westphalia.

The absolute highlight for us students in the early semesters was Professor Immelmann's introductory lecture series on animal behaviour. It was said that he practised his talks at home in front of the mirror before giving them at the lecture hall – whether or not this was true, he was a gifted and captivating speaker, one who could nearly talk in paragraphs. His intro series presented the state of current behavioural biology at the time. While all the topics were fascinating, there was one area of research that interested me above all: density stress in humans and animals.

Here was a world of general rules which seemed to apply equally to humans and other mammals alike. We learned, for instance, that if a number of individuals in a population increases as available space becomes scarce, then the individuals experience stress that is reflected in their behaviour, physiology, reproduction, and health. Studies on mice, rats, and rabbits have shown that when population density increases, individuals become more aggressive towards one another, and mothers take less care of their young. At the same time, production of stress hormones increases, which can lead to health problems and death within the population. Reproductive disorders also occur in parallel, drastically reducing the birth rate and ultimately causing the entire population to collapse. Professor Immelmann presented us with studies of humans alongside those of animals, which showed very similar findings – the same effects have been observed in the satellite cities of metropolises, for example, where the number of inhabitants had steadily increased.

I got my hands on the original articles to which Professor Immelmann had referred and began to work my way into this field of research. Soon I wanted to conduct my own studies. And I had an idea: Immelmann's department kept numerous domestic guinea pigs for study, but nothing was yet known in the literature about how increasing density leads to behavioural changes in these animals. Why not investigate density stress in guinea pigs? Fortunately, Hubert Hendrichs, another professor in the Bielefeld behavioural research department, helped me to begin studying this topic in my fifth semester.

To begin, we put several male and female guinea pigs in a large, semi-covered outdoor enclosure. We always provided the animals with water and food in sufficient quantities, and regularly gave them apples, carrots, and hay between meals. As expected, the population multiplied vigorously: after about a year, there were more than 50 guinea pigs in the enclosure. What gave us pause, however, was that the animals did not behave at all according to the current literature. Our general impression was that the more animals there were, the more comfortable they felt. There were no signs of stress or increased aggression. I wondered what was so different about guinea pigs compared to the species that had been previously studied. Why were they able to handle high density so seemingly effortlessly? Why did they not experience density stress? This investigation brought me to the heart of behavioural biology and my first scientific question as a researcher.

THE SOCIAL INTELLIGENCE OF DOMESTIC GUINEA PIGS

While researching my doctoral thesis, I found the answer: guinea pigs have the amazing ability to form two different types of social organisation, one at low population densities and another at higher ones. By switching in this way, they manage to avoid the density stress normally associated with increases in population size. Let's take a closer look at these social processes.

If a small number of animals – for example, three males and three females – are introduced into an enclosure, the males initially exhibit threatening and aggressive behaviour. After a short time, however, the balance of power is clarified and a linear dominance hierarchy is established. From then on, whenever the highest-ranking animal approaches any lower-ranking ones, they avoid each other. Whenever the second-ranked animal comes too close to the lowest-ranking one, the latter quickly moves out of the way. As a result, most conflicts resolve without escalation: threatening behaviour rarely occurs, and fighting almost never happens.

As one might expect, the highest-ranking male has preferential access to the important resources in the enclosure. He occupies the best hut and initiates courtship and sexual behaviour towards the females much more quickly than his opponents. If any of his rivals approach the females with sexual intent, he attacks them immediately. As a result, the highest-ranking male will usually be the father of the offspring in the colony.

Although much less conspicuous, the females in such a group constellation also form a long-term, linear dominance hierarchy, which manifests itself only in the fact that certain females tend to regularly avoid one another. Their reproductive success is not determined by their position as it is with males: all females give birth to young at regular intervals, who then grow up, and, after reaching sexual maturity, integrate into the existing dominance hierarchy of the population.

When the number of sexually mature animals grows to a dozen or more, a change in social organisation occurs within about four weeks.

The linear dominance hierarchy of the males is replaced by a much more complex social pattern. Studies of colonies of up to 50 guinea pigs show that the group splits into several stable sub-groups, each consisting of one to five males and one to seven females. Each sub-group chooses a section of the habitat towards which its members retreat, especially during resting periods. The males of each sub-group, in turn, organise themselves into linear dominance hierarchies, with the highest-ranking male of each referred to as the alpha. Alphas form strong social bonds with the females of their sub-groups, which can last for years. They almost exclusively look after these females, and will only perform a rumba, their typical courtship ritual (so-called because of its resemblance to the Latin-American dance of the same name), with them. The alphas also guard and defend the females of their sub-group, especially during the mating season, and are the fathers of almost all their offspring.

An astonishing mechanism regulates the relationship between the different alphas: they acknowledge each other's relationships with the females of their respective sub-groups and ignore the females from other groups even when they are in close proximity and ready to mate. The lower-ranking non-alpha males also primarily have contact with the females in their sub-group, but as soon as they begin intense courtship behaviour they are attacked by the alpha. Still, it pays for the non-alphas to form bonds with the females of their group through permanent courtship, since this is the path that leads to alpha status in the long run. Perhaps a previous alpha gets older and weaker and can no longer defend all his females; perhaps one of the females switches her preference from an alpha to a non-alpha and thus turns the lower-ranking male into an alpha within a few days. Surprising as it may seem, in large populations, male guinea pigs achieve alpha status not by fighting for it, but by forming bonds with individual females and investing in them permanently.

Females in large populations also form linear dominance hierarchies within their sub-group, although fighting never occurs and light threatening behaviour only rarely. Females generally like to form social bonds with the alpha males, but it has also been observed that some prefer lower-ranking non-alpha males.

In summary, the social organisation of guinea pigs in large populations can be characterised by three features: First, facilitated social and spatial orientation: by forming permanent bonds and splitting the colony into stable sub-groups, the animals organise social life into clearly manageable units, regardless of whether there are 20, 50, or more individuals in the colony. Second, relative peace: because alphas respect each other's social bonds, they do not compete for the same females. Thus, there is no reason for escalated confrontation and hardly any fights take place. Lastly, high social stability: the positions that individual animals take are constant over many months, and the basic pattern of the social structure exists independently of the animals who occupy it.

Thus, while a linear dominance hierarchy generally characterises small groups of guinea pigs, a far more complex structure can be found in larger ones. The ability to switch from one form of social organisation to another allows these animals to perfectly adapt to increasing population density as social life can be characterised by a clear orientation, little aggression, and high stability no matter the size of the group. Indeed, it is this ability to socially organise that allows these animals to generally remain calm and relaxed under conditions that might otherwise cause density stress.

HOW HORMONES COME INTO PLAY

In retrospect, our research team had quite a stroke of luck: just as we were deciphering the social organisation of domestic guinea pigs, Ekkehard Pröve, another scientist in Immelmann's department, had learned a completely new method for detecting hormones in a laboratory in the US and brought it to Bielefeld. The key feature of this method was that it only required minute amounts of blood to analyse the concentration of certain hormones – including cortisol – present in a sample.

Stress causes the release of cortisol from a hormone gland known as the adrenal cortex. In principle, this is a sensible response since cortisol triggers processes in humans and other mammals that provide the body with energy and resilience, enabling them to adapt to stress. But if the hormone is released too often, in the long term it can have a number of disastrous effects, depleting the organism's energy reserves, weakening its immune system, increasing its susceptibility to disease, and sometimes causing breakdown or death. Cortisol, like adrenaline, is secreted in stressful situations, and both are also known as stress hormones.

This new technique of detecting hormones from only a few drops of blood offered us the fantastic opportunity to combine our observations of guinea pigs with studies of stress hormones to answer fundamental questions about the relationship between stress and behaviour. Is there a relationship between social organisation and stress? Between stress and social status? Are dominant animals less stressed than inferior ones? Do social experiences influence an animal's behaviour and stress response? Can the presence of an animal's bonding partner reduce stress?

But before we could start our research, there was a problem: how could we get blood samples from guinea pigs? Veterinarians told us that the common method was to insert a needle directly into the heart or the blood vessel at the eye. These are serious procedures that were out of the question for us – after all, if we took a blood sample from an alpha male in this way, he would hardly be able to remain in his high-ranking position. The solution came from a nurse who suggested taking the sample in much the same way that he did with his patients: apply a bit of ointment that stimulates blood-flow to the earlobe, then briefly puncture the vessel with a needle. This indeed worked for our guinea pigs, even without ointment, and by now it is a widely practised method of obtaining small amounts of blood from such animals.

Our research confirmed what we had previously concluded from behavioural studies alone: living in large colonies with many animals in a confined space does not seem to cause domestic guinea pigs very much stress. Those living at high population densities did not exhibit higher average cortisol levels than those kept in small groups or in pairs of one male and one female. Interestingly, the hormone levels of the animals also did not differ with respect to rank: the inferior animals, who always had to yield to the dominant ones, were apparently no more stressed than the alphas. Thus, lower status did not necessarily bring about higher physical or psychological stress than higher status.

The prerequisite for this finding was, however, that the relationships between the animals had to be clarified. If the hierarchy was not yet decided and there was frequent friction in the enclosure, then the guinea pigs' stress levels increased. For example, if a non-alpha male tried to overthrow the alpha in his sub-group, both opponents exhibited strong stress reactions that continued until the dominance relationship was established. If both animals arrived at a social position that they accepted, they were able to live stress-free again and their cortisol levels returned to normal. This occurred regardless of whether the original alpha had stabilised his position through the confrontation or there had been a change in the hierarchy. Clear social relationships leading to predictable behaviour on the part of all group members was the basis for this finding.

Still, the question was: Why can domestic guinea pigs, in contrast to many other animals, form these long-term stable relationships? Why are they so adept at organising themselves socially? At first, we thought this question to be trivial – these are not wild animals, and in the course of domestication they must have just been bred to be agreeable. Indeed, when we compare domestic animals to their wild ancestral forms – dogs with wolves, cats with wild cats, horses with wild horses, or domestic guinea pigs with wild ones – the domestic forms prove to be much more agreeable and significantly less aggressive. We were greatly astonished when we learned that this explanation alone was insufficient: guinea pigs must also have specific social experiences during their lives to learn this kind of tolerant and stress-free interaction with conspecifics.

This was particularly evident when adult males tried to integrate into foreign social groups. Only those that were raised in larger mixed-sex groups were able to do this with ease. On the newcomers' first day in the unfamiliar colony, they would explore the new environment and get acquainted with the other inhabitants by sniffing them. They did not attack any males or court any females, and, over the next few days, they assimilated into the existing social structure without any major confrontations, sometimes even taking a higher position than they had in their old colony. Cortisol measurements during this integration phase showed no increase, either in the first hours or in the following days. The animals also did not lose any weight – they were able to integrate into a completely foreign social group of other guinea pigs with little-to-no stress or aggression.

Males that had grown up on their own or with only one female, on the other hand, reacted quite differently: as soon as they met a female in the foreign colony, they aggressively courted her, and as soon as they met another male, they attacked. But in the course of the day, they were defeated by the resident alpha males and retreated into a corner of the enclosure, where they avoided contact from then on and were left alone by the others. These newcomers exhibited a strong stress reaction: their cortisol levels increased almost threefold in the first five hours and only normalised after three weeks. By the third day they had lost 10 per cent of their body weight.

A number of studies have investigated the reasons behind these differences in behaviour and associated stress responses. As Chapter 6 will explain, the results point to the crucial role of social experiences during adolescence, the transitional period between childhood and adulthood. During this phase, males must learn the ability to interact with strangers without stress or aggression through encounters with older dominant conspecifics. Interestingly, the same finding does not apply to female domestic guinea pigs – they nearly always get along with unknown conspecifics with relative ease, independent of their previous life experiences.

Lastly, in a now classic study, we also determined how acute stress can be effectively buffered in guinea pigs, through the presence of a social bonding partner. This can be well illustrated by the story of Ginger Boris, a guinea pig who has since become a television and internet star. Ginger Boris lived as an alpha male in a large colony. When we took him out of it and put him alone in a new enclosure, he quickly began to exhibit stress symptoms, as all animals do in new situations. Within one to two hours, his blood cortisol levels had increased by about 80 per cent. After a few hours had passed and his hormone levels returned to baseline, we returned Boris to his colony. A week later, we reintroduced him to the same foreign enclosure, but this time with a female from his previous colony who belonged to a different sub-group. His cortisol levels rose even more, and climbed still higher when he was introduced to a completely unknown female in the new enclosure a week later. However, when he was introduced to the new enclosure and found his favourite female from his sub-group there, his cortisol did not rise as much as it had in the other situations. Boris's hormonal reaction in an acutely stressful situation could thus be buffered by the presence of a known conspecific.

We observed this bonding-partner effect not only in Boris, but in all males from the colony whom we tested under the same conditions. The same effect was also observed in the females: if their male bonding partner was present, they were far less stressed in the new situation. Overall, the studies demonstrate that domestic guinea pigs benefit from the comradery of their bonding partner, who, if present in a foreign living situation, significantly reduces the guinea pig's stress.

WHAT TRIGGERS STRESS, WHAT BUFFERS IT

The studies we have so far discussed were conducted on domestic guinea pigs and extended over many years, leading to the fact that these mammals are today considered some of the best studied in terms of the relationships between the social environment, behaviour, and stress. (Indeed, in behavioural biology they are used as a model system for this topic.) However, intensive research focused on similar issues has also been conducted on other mammals, both in captivity and in their natural habitat. By taking a comprehensive look at all these results, we can start to see similarities across all mammals, which we will address in the following section. First, let us turn to the question of which social conditions give rise to stress. Then we'll look at the factors that buffer it.

All mammals that live in groups in their natural habitat tend to form dominance relationships – purely egalitarian societies that lack any social stratification do not exist. (This is true even for species who tend towards social tolerance and cooperation, like African wild dogs or bonobos.) When social relationships are clarified, a stable system emerges that benefits all individuals in a group, regardless of their position. This is because all animals can thrive within a stable framework of clear and reliable dominance relationships, meaning an individual in such a society is not stressed by a larger population or its own potential lower status within it. Countless studies on a wide variety of mammals have demonstrated this fact. But why is it so?

A key insight of modern stress research is that if an animal can control or predict the negative consequences of certain behaviours, then their consequences will be far less severe. In stable social environments with clear relationships, high-status animals control a large portion of the social interactions. When an inferior animal approaches a dominant one, for example, a brief threat is usually enough to cause the inferior to back away. But inferior animals have also learned through experience how encounters with other group members will play out. Thus, all animals in the group start to develop certain expectations: 'If I am dominant, the other animal will get out of my way. If the other is dominant, I will get out of his way and nothing will happen to me'; or, 'If another male approaches my females with sexual intent, I will attack him. If I make advances on one of the other male's females, he will attack me. If I signal no sexual interest, things will remain peaceful.' As long as everyday social life operates as expected, all the animals in the society do well. Although high-status animals maintain more control over the others, this does not necessarily cause less social stress. Rather, predictability seems to be the essential condition for a stable social system in which all animals thrive.

Indeed, groups marked by instability and unresolved social relationships paint quite a different picture. Such unpredictable conditions have a negative impact on well-being, triggering strong stress reactions in the animals that can ultimately lead to illness or even death. The crucial question is, then: What causes social instability? And why are some animals unable to form stable social relationships with conspecifics?

THE DEVASTATING EFFECTS OF SOCIAL INSTABILITY

Social instability especially occurs in the wild during the mating season, which is often marked by a high level of social stress. Red deer provide a good example here: during most of the year, red deer stags live together peacefully in what are known as 'bachelor groups'. But when mating season arrives, they become extremely quarrelsome, competing for females through duels and display behaviour. If a dominant male is not clearly decided through these competitions alone, then the stags fight. Fights are accompanied by strong hormonal stress reactions, which cause the animals to lose up to 20 per cent of their body weight, and injuries – and even fatalities – from the clashes are not uncommon.

Quite similar cases have also been observed with wild rabbits. A group of researchers studied a large population of these animals a few years ago on the island of Sylt, first during their mating season in March when aggression was at its highest, then in October/November when the season was over. Both males and females showed a strong release of stress hormones during breeding time.

The most devastating known effects of mating season social instability across the entire animal kingdom were described by the Australian bioscientist Adrian Bradly and his team in their study of marsupial mice: tiny, greyish-brown predators who live in the forests of eastern Australia. Before their two-week reproductive period in the Australian winter, there are about equal numbers of males and females in their colonies; after, only females remain. The males do not survive the mating season. The population is continued only by the pregnant females.

Let's take a closer look at this devastating process: after a gestation period of about four weeks, female marsupial mice give birth to their young and begin nursing them in the early spring. After weaning, the male offspring are driven away to shared nests where they live peacefully until reaching sexual maturity, when they begin to establish territories that they occupy individually and whose boundaries they defend. At this point, the mice live in a well-ordered, stable social system with clear rules.

At the onset of the mating season, however, these territorial boundaries begin to break down. From then on, the males engage almost exclusively in mating with females and fighting with other males. Studies of their stress hormones show them to be at extremely high levels, which, in combination with their elevated sex hormones, leads to a disastrous weakening of the immune system. The male mice are thus defenceless against pathogens, and they perish within a very short time. Importantly, if the males are captured from their natural habitat before the start of the reproductive phase and housed in enclosures that shield them from this social instability, then they do not exhibit these associated stress reactions and can go on to live for several years. Thus, their deaths are not genetically programmed: rather, adult males die because of the social processes associated with their mating seasons.

Why do these marsupial mice subject themselves to such extreme stress? Why don't they simply choose a longer, stress-free life over the chance to mate? An animal that behaved in this way would indeed live a long and healthy life. But such a behavioural pattern would not be, as they say, 'evolutionarily stable', because this animal would have no opportunity to pass on its genes to the next generation. Only members of its species that invest all their energy in reproduction are able to pass on their genes, even if reproduction is extremely stressful and shortens their lifespan. Consequently, the next generation consists only of males carrying the gene for this self-destructive behaviour.

Even animals in captivity are often confronted with socially unstable conditions. This happens particularly when the composition of a group constantly changes: the animals exhibit aggressive behaviour more frequently, and are therefore unable to form lasting dominance relationships and social bonds. This phenomenon can be observed in pets, laboratory animals, farm, and zoo animals alike. These situations are invariably accompanied by a strong release of stress hormones that lead to long-term adverse health effects. Remarkably, the dominant animals are often more affected than the inferior ones, as they must constantly negotiate and reaffirm their high status.

A study by an American research team led by Jay Kaplan on crab-eating macaques demonstrates this process clearly. These monkeys thrive under the same conditions in captivity as they do in their native forests of Southeast Asia, where they live in groups with several adult males and females, and form stable sex-separated dominance hierarchies. When they live under such stable conditions, a diet of high-cholesterol food has no effect on their health. This is surprising, since high cholesterol is known as a risk factor for cardiovascular disease in humans and, according to veterinarians, the same is true for crab-eating macaques. But if animals are repeatedly removed from the group and new ones are added, the resulting social instability causes the dominant males in particular to react with stress and aggression. Thus, while high levels of cholesterol alone have no negative consequences on the animals' health, the combination of cholesterol and social instability poses a significant risk, particularly for the higher-ranking monkeys.

WHY CERTAIN ANIMALS CANNOT LIVE AMONG
EACH OTHER STRESS-FREE

When two unfamiliar males of any species meet, they often engage in a fierce confrontation, which gives rise to a sharp increase in their stress hormones. Whether they then form a stable dominance relationship or reach no amicable agreement and continue to live at a permanently elevated stress level largely depends on two factors: the previous social experience of those particular animals and the type of social organisation that is typical for their species.

In terms of experiences, the studies of domestic guinea pigs have already shown us how the social conditions not only have an effect on the animals' behaviour, but also their stress reactions and state of mind. It is quite fascinating to see how two males who are complete strangers can negotiate their first encounter without any aggression, as long as they have been socialised in large, mixed-sex groups. Males who are socialised without the presence of older dominants, meanwhile, come into these encounters with high levels of aggression, leading to permanent incompatibility and extreme stress.

Similar rules governing the relationship between experience, aggression, and stress are likely to apply to almost all mammals that live in groups of several males and females in their natural habitat. Thus, encountering unfamiliar conspecifics does not necessarily lead to aggression and stress: the rules for a compatible and stress-free coexistence can be learned.

The situation is different, however, for animals from solitary species. The field hamster, for example, lives and defends its territory alone in its natural habitat – thus it would not be a good idea to house two adults of this species together in captivity. They would not come to terms with one another; their stress levels would remain extremely high. Such reactions have been described in many other solitary and territorial species as well.

But while peaceful coexistence can in many cases be learned, not all mammalian species that usually experience stress when encountering same-sex conspecifics in the wild can achieve it. Tree shrews have been particularly well studied in this regard thanks to the work of the zoologist Dietrich von Holst and his team at the University of Bayreuth. These animals bear some resemblance to squirrels in size, shape, and appearance, although their heads are pointed and their mouths are studded with sharp teeth. But contrary to what their name may suggest, they are not rodents – they are a distinct group of mammals closely related to monkeys. Tree shrews live in pairs in the tropical and subtropical forests of Southeast Asia and tend to mark and vehemently defend their territory against other conspecifics. In captivity, when two unfamiliar males of this species are introduced into an unfamiliar enclosure that contains several retreats, sleeping boxes, and feeding and drinking sites, they explore the new environment at first. Within the next few hours, they begin to confront each other, and over the course of one to three days a winner is decided. As expected, during the first phase of confrontation, both animals experience a sharp increase in stress hormones and heart rate. After the dominance relationship is established, however, the dominant hardly pays attention to the inferior, and aggressive confrontations are very rare if present at all. As a result, the stress level of the dominant animal returns to normal.

For the inferior, however, the situation is quite different. Scientific observation divides their behaviour into two categories: One type crawls into a corner of the enclosure or retreats into a sleeping box, which he only leaves to hastily eat or drink. He loses initiative, stops grooming himself, and appears apathetic and depressed. Such a behavioural pattern is known as 'passive coping' or 'passive stress' in humans and animals and is accompanied by extremely high cortisol levels, which can lead to a serious weakening of the immune system within a short time.

The other type, meanwhile, reacts to his loss in almost the exact opposite way: he becomes hyperactive and hectic, always watching out for the dominant and trying to get out of his way. If this fails, he actively defends himself despite his inferior position. Such a pattern is called 'active coping' and is accompanied by a strong release of adrenaline and noradrenaline, as well as a permanently increased heart rate.

Thus, there exists no possibility of stress-free coexistence when two same-sex tree shrews meet. The struggle between the two males awards the dominant with a stress-free life, while the inferior is caught between a rock and a hard place, so to speak, reacting with either active or passive coping to the stress of the situation.

Nonetheless, we cannot generalise from the example of the tree shrew that the dominant animals are always the ones who get to live stress-free. The crab-eating macaques have already shown us that it is sometimes the animals of higher status who have particularly violent stress reactions to socially unstable conditions; in domestic guinea pigs, we observed that stress levels rose in both alphas and non-alphas when their relationships were unresolved. Thus, it is not the status of the animal *per se* that determines whether it experiences an increase in social stress, but rather the behaviour associated with its status. In tree shrews, the winner pays no attention to the loser and is therefore able to live without stress. In many monkey and mouse societies, on the other hand, the highest-ranking males must constantly reaffirm their dominance by guarding the females, exhibiting threatening behaviour, or patrolling territorial boundaries. These animals must therefore act like 'managers' and are subject to a high level of active stress. Behavioural biology refers to the stress incurred in such cases as 'the cost of dominance'. As has been very well studied in mice, these costs are associated in the long term with increased blood pressure and other cardiovascular diseases.

THE BLESSING OF GOOD RELATIONSHIPS

For many years, stress research tended to focus primarily on identifying causes for different animals. Gradually, however, scientists increasingly began to also ask what alleviates stress. Domestic guinea pigs have given us one important insight into the role of the bonding partner in reducing hormonal stress reactions. But this does not only hold true for guinea pigs: the proximity of conspecifics in general has proven to be one of the best remedies for stress, provided the animals in question have good social relationships.

Almost all mammals exhibit close bonds between the mother and her offspring, especially while she is nursing her young. During this phase, the mother is not only important because she feeds, warms, and protects her children, but also because she knows how to keep their stress levels down in exciting situations. Whether we look at guinea pigs, rhesus monkeys, squirrel monkeys, or humans, we see that young children experience a rapid increase in cortisol when they find themselves without their mother in a new situation. If the mother is present, however, the children exhibit no stress response. Interestingly, in non-human mammals, it is by no means always the mother who is best able to buffer her offspring's stress response – for example, in the coppery titi, a species of New World monkey, the father does this much more effectively.

The reason behind this becomes apparent when we compare coppery titis to squirrel monkeys. Both are species of New World monkeys that live in the forests of South America; they are approximately the same size and eat roughly the same things. But the social organisation of each species is completely different: squirrel monkeys live in large, roving, mixed-sex groups in which the members are most likely to associate with their own sex. Coppery titis, on the other hand, form monogamous pairs who live together with their offspring in territories that they vehemently defend against other conspecifics. These different lifestyles have a clear impact on the parent–child relationship – squirrel monkeys exhibit a close emotional bond between the child and its mother, and encounters between the father and its offspring are rare. It is therefore unsurprising that the mothers of this species are much better able to alleviate their children's stress than the fathers are.

In the coppery titis, by contrast, both parents are involved in raising the young (as is the case with many other species who live in pairs), but the main job of the caretaker is taken on by the father, who carries the child around and often only leaves it with its mother to nurse. It follows, then, that the offspring bond much more intensely with the father than with the mother, and the father is thus better able to manage the stress levels of his young. The general rule that emerges from these studies is that the closer the animals' social relationship, the better their ability to protect against stress.

Ginger Boris has already shown us how this rule also applies to adult relationships. It can also be deduced from the behaviour of squirrel monkeys and coppery titis, as the Californian psychologists Sally Mendoza and William Mason were able to demonstrate: monogamous titis form close emotional bonds with their partners and are able to keep each other's stress levels low. When a titi loses a partner, its stress levels rise immensely. In squirrel monkeys, by contrast, such an emotional bond between males and females does not exist, either in their natural habitat or in captivity. Accordingly, these monkeys do not experience much stress when males and females in their group are separated, even if they had lived together as a pair for a long while. In stressful situations, however, the presence of a mate does not have a calming influence.

It is not only partners of the opposite sex who are able to reduce stress: same-sex social partners can buffer it equally well. In numerous monkey species, for example, females are part of a social network of closely related conspecifics with whom they form powerful bonds. Here, too, it has been shown time and again that the stronger the bonds and the closer the network, the lower the adverse reactions of the members when plunged into a stressful situation.

Finally, social bonds between males can also effectively buffer stress, as studies by a team of primatologists in Göttingen, led by Julia Ostner and Oliver Schülke, on Barbary macaques demonstrates particularly well. Barbary macaques live in groups of several males and females. While it is true that the males engage in fierce competition over the females, they also form close social relationships with each other, which are manifested by spatial proximity, frequent physical contact, and grooming of each other's fur. These relationships have been compared to friendships between humans and provide an effective buffer against the stress of everyday life. In these monkeys' natural habitat in the Moroccan Atlas Mountains, these stressors mainly include aggression from other conspecifics and frigid temperatures, both of which can lead to a permanent increase of stress hormones. Interestingly, the more developed the relationships these animals have with other males, the lower their adverse reactions to stress from both the social environment and the weather.

Some of the most astounding findings concerning social relationships come from tree shrews. As we have already discussed, these animals live in pairs in their natural habitat. However, when a male and a female are first introduced to each other in captivity, in most cases they do not form a pair – indeed, sometimes they engage in such intense fighting that they must be separated. Usually the two can coexist, but there remains a high level of tension between them that manifests in mutual avoidance and occasional confrontation. In about 20 per cent of these encounters, however, something completely different occurs: the animals meet, they take a liking to each other, and they immediately begin a friendly relationship. One is given the impression of 'love at first sight': the animals lick each other's mouths, are constantly near each other, and always cuddle when they rest. They always sleep in the same box at night, which pairs who do not get along never do. If we compare the stress levels across different couples, a clear difference emerges: the harmonious couples are drastically calmer. As a result, they have much better immune systems and permanently lower heartbeats than discordant couples. As my mentor Dietrich von Holst used to put it in his lectures on tree shrews: 'Sometimes love is the best medicine for animals, too'.

CONCLUSION

This chapter has focused on the fundamental importance of the social environment to mammalian behaviour and stress levels. We have seen how social contact can have both positive and negative effects: if animals are integrated into a stable social system in which each individual knows and accepts its position, then neither low social status nor high population density necessarily increases their stress levels. But under socially unstable conditions in which relationships are not clear, certain animals may experience strong stress reactions and adverse health effects. Whether stable relationships can be established depends largely on the social organisation of the species, as well as the previous social experience of the individuals. Being part of a strong network and having good relationships with bonding partners have proven to be particularly effective protection against stress.

CHAPTER 3

Cats Are Happy When They're Playing

On the Well-Being, Emotions, and Humane Treatment of Animals

EMOTIONS AND ANIMAL WELFARE: LONG-NEGLECTED TOPICS OF BEHAVIOURAL BIOLOGY

HARDLY A DAY GOES BY WITHOUT A NEWS REPORT ON THE debate over the ethical treatment of animals. Society is increasingly asking questions about what humane treatment looks like: what is the ideal environment for an egg-laying hen? How do polar bears and tigers feel in captivity? Should horses be kept alone in stables, or with other horses? How does a pug feel when its owner goes on holiday and it is left alone?

When we keep animals in captivity, we are responsible for treating them well. But how do we know when they're thriving or suffering, or when it is appropriate to keep them at all? European Union legislation states that keepers of animals must avoid causing them pain, suffering, or harm. But an animal's individual experience is in many ways subjective, and is not easily identified through scientific procedures. There is therefore a growing demand for research into the proper conditions for humane treatment to focus on animal emotions. The experience of positive emotions is increasingly regarded as a central aspect of their well-being.

Behavioural biology is thus faced with the task of developing scientific criteria and methods for evaluating animal welfare and emotions. It is certainly not enough to simply look at animals and make subjective judgements from their appearance – dolphins, for instance, always appear to be laughing, but this impression is only created by the shape of their jaw and their relative lack of facial expression. We cannot take it to mean that they are constantly in a good mood and are thriving.

The task is still more challenging since the founding fathers of behavioural biology omitted the topics of emotions and well-being from their work. Such experts as Konrad Lorenz undoubtedly knew that animals have emotions, but he maintained that it is not possible to make scientifically sound statements about their subjective experiences. In hindsight, his avoidance of the topic was also likely a strategic move in the early days of behavioural biology as it was difficult enough to establish animal behaviour as a serious field of research and convince the scientific community of the merits of objectively describing behaviour. Had these pioneers also taken emotions into account, it would have been virtually impossible to establish behavioural biology as an independent discipline more than half a century ago; consequently, animal welfare and emotions were overlooked for decades. Recently, however, this has changed: scientists have developed methods to assess these qualities and identify the factors that can lead animals to suffer and those that can help them thrive.

Welfare is a mutable state in humans and animals, one that can range from very good to very poor. We cannot, of course, assess animal welfare with something as straightforward as a questionnaire, but in many cases poor welfare is relatively easy to detect. When factory-farm pigs begin to chew off each other's tails, chickens start to peck off each other's feathers, and cattle break their limbs during transit, for instance, it does not require much research to declare that the welfare of these animals is under threat. It is also relatively easy for veterinarians to identify afflictions caused by infections, parasites, or tumours associated with severely impaired health. But if none of these physical signs are apparent – if a horse, pig, dog, cat, guinea pig, mouse, or parrot looks fine at first glance – how do we know if they really are? If we detect no afflictions, can we conclude that an animal is in excellent health?

Most behavioural scientists would agree that we probably cannot. So how does current behavioural research go about performing a welfare diagnostic? Essentially, scientists need to assess both the physical and mental health of an animal. Physical health obviously includes the absence of disease and bodily damage in accordance with the typical life expectancy of the species, while psychological health is assessed partly through other physiological measurements, such as the concentration of stress hormones, and partly through observation of an animal's behaviour.

HORMONES AND ANIMAL WELFARE

We have already explored how measuring certain hormones allows us to make claims about the relationship between environment, behaviour, and stress. In general, such measurements can be used to assess whether an animal is able to adapt to captivity, or whether its enclosure, its social partner, its caregiver, or indeed its very existence under human care represents an excessive demand. Such hormone tests often lead to insights that we would not gain from observing behaviour alone.

When a guinea pig, for example, is removed from its group, carefully placed on a human's lap and petted for 10 minutes, it sits quietly and seems content. All signs seem to indicate that it is doing well. But if we take a saliva sample from the animal's mouth at the beginning and end of the petting session and test it for cortisol levels, we see a different picture: the animal's stress hormones show an increase of almost 80 per cent. Clearly this guinea pig does not particularly like being stroked, at least if it is not used to people. If, on the other hand, it is only briefly removed from the enclosure for a sample to be taken and then immediately returned to its familiar conspecifics, a sample from 10 minutes later shows no increase in stress hormones.

As a rule, however, hormone levels alone do not provide insight into an animal's welfare – rather, conclusions can only be drawn from a combination of hormone measurements and behavioural observation. An experience my colleagues and I had at the Allwetterzoo in Münster illustrated this point nicely: a few years ago, we had found that the way in which a group of white rhinos was fed had a dramatic effect on their stress level. At the time, a bull called Josef lived with four females – Natala, Emily, Vicky, and Emmi – in an outdoor enclosure during the day. The rhinos spent evenings and nights in individual pens where they received their main meals. In the mornings they came out to the larger enclosure together, where they would always gather around a fresh pile of hay and eat it up within a half hour. What worried us was the relatively high level of aggression the rhinos would then exhibit throughout the day.

We wondered if perhaps the morning hay was responsible for the aggression, so we conducted a study in which we systematically varied the rhinos' feeding method: they would always receive the same amount of hay in the morning, but on some days we would provide it in one large pile in the centre of the enclosure, and on others in five small mounds evenly distributed throughout it. We had also just recently developed a method of using saliva to determine an animal's concentration of stress hormones, and therefore took samples from the rhinos every morning and evening. Remarkably, the animals' stress levels were significantly elevated in the evenings after they had found the hay in one large pile – these changes could even be detected the next morning.

But why did this method of feeding still result in increased stress 24 hours after feeding? Behavioural observations provided the answer: when eating from a single haystack, all five of the rhinos came quite close together. The proximity triggered aggression, especially between the male and the females, which was maintained throughout the day. When feeding was done in five piles, the animals did not come nearly as close to each other, meaning that their aggression and stress levels remained significantly lower.

Stress hormones like cortisol, corticosterone, and adrenaline are present in comparable forms in all vertebrates, including humans. The amount of stress hormone an individual secretes is an important indication of its stress level, and measuring such hormones accordingly plays an important role in animal-welfare research. Other important physiological measurements include an animal's heart rate, which, when excessively high and irregular, can indicate stress, and its immune system, which stress can also degrade in the long term. But an important finding of the last years has also been that physiological indicators alone lead to a welfare diagnosis that is highly susceptible to error. It is therefore always essential to also record an animal's behaviour.

It is easy to explain why this is: social instability, defeat in decisive confrontations, or separation from a social bonding partner all lead to a powerful release of stress hormones in many animals. It thus follows that such situations should therefore be avoided for animals in captivity. But the release of stress hormones alone is not a sufficient indicator that an animal is suffering: mating, for instance, regularly leads to one of the strongest stress responses that an animal may experience, yet there is nothing to indicate that it is associated with impaired welfare – quite the opposite. The release of stress hormones should therefore only be viewed in connection with an animal's behaviour. In this context, it is important to understand what the primary function of stress hormones is: to trigger processes that supply an organism with energy so that it can, depending on the situation, fight with a rival, flee from danger, or mate with a partner. Many of the situations associated with a stress response are indeed threats to an animal's welfare, but not all.

BEHAVIOUR AND ANIMAL WELFARE

What can an animal's behaviour tell us about its welfare? What kind of activity is a sign of general health and what kind is a sign of suffering? When any animal, be it a dog, a cat, or a hamster, is not eating or drinking enough when food and water are sufficiently available, their welfare is most likely impaired. When an animal begins to neglect its personal hygiene or languish in the corner of its enclosure, its welfare is likely severely degraded. Stables, enclosures, and cages in which such behaviour regularly occurs are certainly not humane.

Another good indicator of impaired welfare is the daily cycle of an animal's behaviour. All species behave according to a particular rhythm of fixed phases of rest and activity. Most songbirds, for example, are diurnal, and exhibit their main behavioural maximum in the early morning and their secondary maximum in the evening. During the night they are completely quiet. Hedgehogs, mice, or hamsters, on the other hand, are active in the night and rest during the day, while domestic and wild guinea pigs follow a rhythm of several phases of rest and activity, which alternate regularly every few hours regardless of the daylight. No matter what form of circadian cycle a particular species has, if an animal is doing well, it exhibits a regular rhythm. Changes in this rhythm are often the first signs that something is wrong, and when these cycles break down completely the animal's welfare is evidently in severe peril.

More subtle signs of impaired welfare can be seen in what are called conflict behaviours. As in humans, it is by no means always clear in animals which behaviour should be performed next. If an animal tries to satisfy two incompatible behavioural tendencies that are activated to approximately the same extent at once, a completely nonsensical behaviour may emerge. We can see this, for example, in fighting roosters, who in the midst of a fierce confrontation will sometimes abruptly stop and start to peck at imaginary grain. Something comparable also occurs with oystercatchers, who will suddenly assume a sleeping position during a fight, only to quickly continue again as if nothing had happened. Such unexpected behaviours that occur out of context are called displacement activities and indicate that an animal is experiencing some sort of inner conflict. Thus, if the animals in enclosure A are performing significantly fewer displacement activities than those in enclosure B, we can deduce that that the animals in the first enclosure are doing better than those in the second.

Another form of conflict behaviour is known as vacuum activity, which occurs when a behaviour breaks free of its external stimuli and runs completely of its own accord. Weaver birds, for instance, build very elaborate nests of grass in the wild. But if they are kept in aviaries without nesting material, they nevertheless perform very complicated constructing movements while idling and appear to the human observer as if they are building imaginary nests. The occurrence of such vacuum activities indicates that certain behavioural systems like nest building have been strongly activated in the birds but cannot be meaningfully executed. Animals in captivity who are frequently found engaging in vacuum activities are not living in a humane environment.

BEHAVIOURAL DISORDERS

The type of conflict behaviour that has been the most researched and discussed in recent years is what is known as stereotypy: the constant, uniform, and apparently functionless repetition of a certain behavioural pattern. Such behaviour is common in farm, zoo, laboratory, and domestic animals. Pigs in industrial farms, for example, may bite the bars of their pen for hours on end; predators in zoos may tread the same path back and forth; lab mice may repeatedly scratch at the walls of their cages. Stereotypies can develop from what is known as appetitive or searching behaviour, which originally serves to find a suitable environment in which to satisfy an urgent need. Animals under restrictive conditions in captivity, however, often cannot succeed in satisfying these needs, and thus their searching behaviour turns rigid and repetitive, becoming a disorder, namely a stereotypy.

Neurological studies in rodents have shown that stereotypies are associated with pathological changes in the brain. Such stereotypical movements are very similar to symptoms of certain human psychiatric or developmental conditions – imagine the perpetual upper-body rocking movements sometimes exhibited by individuals with autism.

However, the cause of an animal's stereotyped behaviour is not always the current conditions in which it is kept. Stereotypies can also be caused by traumatic experiences and cannot be resolved by an improvement in habitat. There is the known case of a polar bear who first developed a stereotypy when living in a cramped circus wagon, and who, even when moved to a spacious outdoor enclosure, continued to walk in rigid paths that matched the dimensions of the old wagon.

Stereotypies are generally classified as behavioural disorders. They stem from present or past mistreatment that an animal has experienced. Changes in how animals are kept that lead to a decrease or disappearance of stereotypies are therefore a step in the right direction towards more humane housing conditions. Stereotypies in polar bears were greatly reduced, for example, when their keepers didn't simply deliver fish at mealtime but gave it to the bears in blocks of ice instead. Mice have stopped performing stereotypic behaviours altogether when they are moved out of unstructured plastic cages to larger ones equipped with houses, climbing racks, and other stimulating objects.

Occasionally it has been argued that while stereotypies may offend the aesthetic sensibilities of humans, there is no evidence that they are behavioural disorders. The best counterargument to this idea comes from a study by the British scientists Ros Clubb and Georgia Mason, who looked at 35 different species of carnivores in zoos and compared the mortality rate of their captivity-born offspring to the size of their foraging area in their natural habitat. The study included species that roam very far in the wild, such as polar bears and lions, but also species who roam a relatively small area, like arctic foxes and the American mink. The analysis showed that the larger the animal's roaming area in the wild, the more frequently it performed stereotypies in captivity, and the more stereotypies an animal performed, the higher the mortality rate of its offspring. These data clearly demonstrated that stereotypies are behavioural disorders, and further raised the troubling question of whether any species of animal can be humanely kept in captivity at all.

POSITIVE EMOTIONS AND PLAY

An animal's behaviour can not only indicate when it is struggling but also when it is doing well. This is, for example, demonstrated by what we call socio-positive behaviour: animals licking, cuddling, and generally behaving nicely towards one another, as we saw in the example of the tree shrews in the previous chapter. We can also infer that an animal is doing particularly well from the types of sounds it makes, something for which the Estonian neuroscientist Jaak Panksepp's 'laughing rats' have become somewhat famous.

Rats, especially when they are young, love to engage in horseplay. When doing so, they emit an abundance of high-pitched whistling sounds that resonate at about 50 kHz – a frequency not audible to humans but detectable by ultrasonic devices. Remarkably, rats also make these sounds when they are tickled by humans, even more so than during play. If a researcher begins to tickle these animals, the rats will often seek out the scientist's hand for more; they will even solve mazes for a tickling reward. The rats who laugh the most when tickled are also the most playful. (As might be expected, they will also abruptly stop laughing when faced with danger, fear, or anxiety.) Laughter and joy, clearly, are not unique to humans.

The close connection that play has to positive emotions is one of the reasons why it is currently a main focus of behavioural biology. The renowned British journal *Current Biology* even devoted a special volume to the topic on its 20th anniversary entitled 'The Biology of Fun'. Animals that engage intensively in play have positive emotions and are obviously doing well; conditions that allow for play are therefore considered appropriate for keeping animals in captivity. Dogs and cats are not the only animals who like to play extensively – apparently all mammals do, as well as many bird species – the New Zealand mountain parrot (or kea), for example, is known as a particular play fanatic. Play is often limited to the young of a species, but it can also be maintained through adulthood, as is the case with many apex predators, primates, whales, and parrots. Play has also been observed in some reptiles, amphibians, and fish, as well as some invertebrates, squids, crested spiders, and field wasps. (Whether invertebrate play is also associated with positive emotions, however, is highly controversial.)

How can we distinguish play from other categories of behaviour? In behavioural biology, play is defined as non-serious behaviour – it has no apparent function in the context in which it occurs. Prey games, for example, are often directed at an object – think of the ball of wool that a cat bats around. During fighting games, dogs or monkeys may reverse roles several times within a short period, sometimes allowing one to be the winner, sometimes the other. This never occurs in serious confrontations. And it is only during play that completely different types of behaviours, such as fighting and sexual activity, can be harmoniously combined. Moreover, play is often marked by exaggeration: many animals often carry out a certain behaviour with greater speed, movement of the extremities, and repetition than when executing it seriously. Play also occurs spontaneously, appears nearly indefatigable, and is regularly sought out. Neurobiological studies suggest that, in vertebrates, play activates the brain's reward centres, and is therefore an activity that is unlikely to come to an end on its own.

Play is also not a one-dimensional phenomenon: there is social play, object play, and what is known as solitary play, during which animals often perform bizarre movements – a guinea pig, for example, will sometimes suddenly run off, stop abruptly, jump, spin, toss its head, and land on the ground again. The animal can repeat this sequence for minutes at a time, and it has a contagious effect on the other guinea pigs nearby, who will occasionally perform serious bouncing fits.

Play is also associated with a high energy expenditure and, in an animal's natural habitat, increased danger. Young animals playing can attract the attention of predators; chase games on a rocky landscape can cause accidents and broken bones. Nevertheless, play occupies a large part of the lives of many animals. Therefore, according to Darwinian logic, it must be beneficial for the individual. And indeed it is: animals can learn through play, which supports the development of muscular, cognitive, and social skills.

Play is characteristic of many juvenile animals, but it does not occur in any given situation. Animals must first be in a relaxed environment, which provides both stimulation and security. If either of these components is missing, play is drastically reduced or does not occur at all. Many animals in human care live in enclosures that are too small and lack any spatial structure or opportunities for stimulation. Under such conditions, animals do not play – on the contrary, they often develop behavioural disorders like stereotypies. The lack of one or more social partners can also lead to under-stimulation, as is especially the case for animals who live in groups in the wild. If the young of such species grow up alone, they play significantly less frequently: young guinea pigs who are raised individually, for example, play much less than those who live in larger colonies.

But a stimulating environment is not the only prerequisite for play – an animal's basic needs must also be met. For instance, the infants of an East African species of long-tailed monkeys are known to play exuberantly and frequently. Play is reduced during droughts, however, because the animals must spend most of their time and energy foraging. Under adverse weather conditions, too, when predators are lurking and social tensions within the group are high, the infants refrain from it. But in a secure and stimulating environment they will engage tirelessly – depending on the species – in solitary, object, or social play.

ENVIRONMENT AND ANIMAL WELFARE

Various studies on farm, zoo, laboratory, and domestic animals have investigated the extent to which environment influences behaviour and welfare. As a general rule, it is clear that animals who are raised in richly structured and varied environments differ markedly from their conspecifics who live in barren and minimally structured ones. The dramatic effects that a rich environment has on laboratory mice, for example, have been well documented.

Lab mice are widely used in biomedical research. Millions of them are studied annually to understand the causes of cancer as well as cardiovascular and neurodegenerative diseases. But these mice do not spend the majority of their lives under experimental conditions – they generally live with other conspecifics in small, rectangular plastic cages that are about 15 cm high and sealed at the top with a gridded lid that contains a trivet for dry food and a water bottle. The approximately 900 cm^2 of floor space in the cage is usually lined with a thin layer of litter. About two decades ago animal advocates first began to criticise these conditions, suggesting that the cages for lab mice be structurally enriched. They proposed that a wooden climbing frame and a plastic insert with openings on all sides through which the mice could climb be added to the cages. Indeed, a number of studies have shown that mice who live in such enriched cages are significantly more active, curious, and confident than their counterparts in standard enclosures, and tend to perform much better in learning tests that present a maze to be solved for a reward.

These results clearly demonstrate that certain structural enrichments have a positive effect on lab mice. Nonetheless, it can still be argued that such enhanced cages are not optimal. A research team I was a part of therefore set out to try and design an environment in which we might like to live if we were mice. The result was what we termed 'super-enriched housing', and our article on the experiment made the cover of a prestigious American journal. What we designed was a 35 cm-high glass terrarium with 4000 cm^2 of floor space. The ground was covered in bedding and scattered with paper towels for nest building. There were also many stimulating objects available: a plastic house, climbing frames, ropes hanging from the ceiling, and a mezzanine level reachable by stairs. Over the course of several-hundred hours, we observed and recorded the behaviour of female mice living in groups of four in either the standard, enriched, or super-enriched enclosures.

Surprisingly, the behaviour of the mice in standard and enriched housing hardly differed. But there were clear differences between the mice living under both these conditions and their conspecifics in the super-enriched environment. The animals in the former conditions frequently exhibited stereotypies, scratching the walls repeatedly with their front paws. They rarely engaged in the running and jumping-style play typical of their species. Aggressive behaviour towards conspecifics was common, and socio-positive behaviour rarely occurred.

In contrast, we observed the exact opposite behavioural profile in the mice from the super-enriched enclosure: they played often, exhibited hardly any stereotypies, were kind to each other, and rarely became aggressive. All animals were of the same sex, age, and genetic makeup, and lived in groups with the same number of conspecifics, but the super-enriched environment resulted in completely altered behaviour and significantly improved welfare.

As numerous neuroscientific studies have shown, the positive effects of an enriched environment can also be observed in the brain. Animals raised in a richly structured environment, for example, generally have a larger cerebral cortex, stronger arborisation, and a higher number of connections between nerve cells compared to conspecifics from more barren environments. An enriched environment even proves to be an effective protection against Alzheimer's disease in mice that have a predisposition to the disease: protein deposits in the brain, which are typical of Alzheimer's disease in humans and animals, form to a much lesser extent in these animals than in those living in standard housing. Their brains also tend to form many more new nerve cells. An environment that promotes an active and varied lifestyle thus appears to benefit humans and animals alike.

ASKING THE ANIMALS THEMSELVES

As we have so far discussed, we can make inferences about an animal's welfare based on observations of behaviour like stereotypies or play, and determine their stress levels by measuring the concentration of certain hormones in their bloodstream. We can also ask the animals themselves, in a sense, to find out what is important to them, what they like, and what they are averse to. The answers that animals give us to these questions are essential to understanding the world from their point of view.

Just a few years ago, it was a relatively widespread practice to keep guinea pigs and dwarf rabbits together – many pet stores even advised customers buying a guinea pig to also buy a dwarf rabbit as a companion.

But dwarf rabbits and guinea pigs are two relatively distant species whose ancestors live in quite different habitats in the wild, which raised legitimate questions as to whether keeping these two species together is actually appropriate. We therefore built an apparatus that contained several compartments in which a guinea pig could choose whether it preferred to live alone, with a dwarf rabbit, or with another guinea pig. Their preference was quite clear: they did not want to be alone or with a dwarf rabbit, but they clearly wanted to live with a conspecific.

The reason for this is also evident in the different species' behaviour: for one, guinea pigs and dwarf rabbits have different rhythms of activity, and the rabbits repeatedly disturb guinea pigs during sleep. Secondly, the two species 'speak a different language', as it were – a dwarf rabbit will exhibit a behaviour called ducking, for instance, in which it slowly moves towards another creature, lowers its upper body and head, pushes its ears back, and places its head on the other's head or chest. This is meant to elicit a positive response, and the rabbit's conspecifics will typically react by grooming, sniffing, or rubbing noses. The guinea pig, however, understands these motions as hostile and will almost always respond with defensiveness. The results of our study clearly demonstrated that guinea pigs should therefore neither be kept alone nor with a dwarf rabbit. The only appropriate arrangement for them is living with another guinea pig.

Such tests of preference in which the animals are offered different lifestyle alternatives have been conducted with a wide variety of species. Mice that are given the choice between standard and enriched housing consistently choose the latter, even if they had previously lived in a standard cage. Tests have also been conducted to determine the ideal flooring for chickens and piglets, the ideal sleeping mats for cattle, the preferred pen temperatures for pigs, and the preferred mating partners for sheep. The findings from these tests should be used more frequently in animal-friendly housing design in the future.

Preference tests do indeed tell us a lot about how animals see the world. However, they do not always indicate how relevant the choice an animal makes is to its general welfare. If a dog has to choose between a bone and a can of food, for instance, it might not always choose the food. But will the dog suffer if it still has to eat the food it did not choose? How can we judge whether an animal's choice is a necessity or a luxury?

Answering these questions requires a method of determining the importance of an animal's preferences. It is reasonable to assume that the more important something is to an animal, the more the animal will be willing to 'work' for it – that is, spend time and energy, take risks, and overcome obstacles to obtain it. How can we take this into consideration in scientific studies?

The British biologist Marian Stamp Dawkins is responsible for a decisive breakthrough in this field. She proposed borrowing a framework from studies in behavioural economics: people with a fixed low income tend to buy a certain amount of bread and a certain amount of champagne in a given period. If both goods become more expensive, they will generally still buy the same quantity of bread no matter the price. Thus, we conclude that the demand for bread is inelastic – it is a necessity. The same cannot be said for champagne: the higher the price, the less is generally bought – it is a luxury. In principle, the relationship between the price of any good and the quantity in which it is purchased can be shown in what is called a 'demand curve'. We can use these curves to determine – in accordance with the definitions that economists have ascribed – whether a certain good is treated more as a necessity or a luxury.

So how can demand curves be applied to animals? Let's take the example of rats, who can learn that pressing a lever will give them a food pellet. We can determine how much food the rat consumes over the course of the day by how many times it presses the lever. We can then change the number of times the rat must press the lever to obtain a single pellet continuously, increasing the ratio to two-, five-, ten-, or twenty-to-one, and once again determine the aggregate amount of food the animal consumes over the day under the changing conditions. Such studies regularly show that, no matter how much effort, time, and energy the animal must expend, it will always procure the same amount of food.

A second step can then similarly elucidate how much work an animal is willing to put in for a resource other than food – for example, a larger enclosure. We can calculate demand curves for both rewards – food and access to space – by corresponding the 'price' of the resource with the effort required to obtain it, in this case the required number of lever presses, and the 'consumption' with the number of rewards the animal ultimately obtains. The curves we draw can then help us conclude an animal's elastic and inelastic demands. The charm of this method lies in the fact that it can be applied in the exact same way to the consumption habits of humans and animals.

Demand curves have been used in a whole series of studies in recent years. As is perhaps expected, animals always have an inelastic demand for food. Pigs kept in individual pens, however, will work almost as hard to interact with a conspecific as they will for access to food, so social contact might also be considered an inelastic need. Similarly, a mouse's demand for additional space and other enrichments and a laying hen's for nesting boxes also prove to be relatively inelastic, and can thus be considered basic needs. We can assume that animal-friendly conditions are those in which all inelastic needs are regularly satisfied. Animals suffer when these demands are not met.

As useful as they are in determining an animal's wishes, preference tests also raise a problem: animals, like humans, do not always choose what is best for themselves in the short term. When given the choice, rats will always vie for treats over more nutritional food. When given the choice between alcohol and water, a large proportion of animals will regularly choose alcohol and even form a dependence on it. Thus, not every experimentally determined preference is beneficial – preference tests should always be supplemented by other forms of study, such as behavioural observations and physiological evaluations.

OPTIMISTS AND PESSIMISTS

An animal's hormone levels can tell us about its stress, its behaviour can key us into its general welfare, and its preferences can help us determine what it deems important. But all these metrics do not help us to answer a more fundamental question: How do animals feel? What emotions do they have? At the start of the millennium, the British behavioural biologist Michael Mendl and his team developed a brilliant way to address this topic scientifically.

The team's starting point was a basic fact of human psychology: how we evaluate the world around us depends to a large extent on how we feel. For example, when happy people are asked about the future, they tend to be optimistic: 'Everything will be fine in the end!' In contrast, unhappy, anxious, or depressed people tend to react pessimistically to the same question, fearing accidents, unemployment, loneliness, or illness. Ambiguous situations are also evaluated quite differently by people in different emotional states, something famously encapsulated in the 'glass-half-full' analogy. A large number of scientific studies have also confirmed this general intuition: emotions influence human thinking in the broadest sense, a phenomenon known in psychology as cognitive distortion.

Mendl's team therefore reasoned that, while it may not be possible to measure an animal's emotions directly, it should be possible to determine its cognitive biases, as one would with humans, and thus infer the emotions behind them. In a ground-breaking experiment with rats, the research team showed that animals can indeed demonstrate whether they are optimists or pessimists. In the first step, the rodents were trained to distinguish between two tones. When the first one played, they were allowed to press a lever and receive a food pellet; the tone thus announced something positive. Pressing the lever when the second tone played produced an unpleasant noise, which the rats were trained to avoid by refraining from pressing the lever. After the rats learned to distinguish between the two sounds and their consequences, the research team asked the exciting question: what would happen if a tone was sounded in the enclosure that was exactly between the two previous frequencies? Would the rats press the lever – that is, react as if they were expecting something positive – or not?

Interestingly, rats that had lived in poor and chaotic housing conditions where their bedding was often damp, or the lights were turned on at irregular intervals, turned out to be significantly more pessimistic than conspecifics who had lived in more stable conditions. In concrete terms, this meant that they took longer to press the lever and pressed it significantly less frequently than animals from the more humane enclosures, and thus interpreted the same tone significantly more pessimistically. Their previous experiences had caused a cognitive distortion, which suggested the animals lived in a more negative emotional state.

In principle, this method hinges on an animal's ability to distinguish between a positive and a negative stimulus. The crucial question that follows is then: How will the animal react when presented with a stimulus that exists exactly between the two it has previously learned? Will the animal turn out to be more of an optimist or a pessimist? Mendl's team used sounds as stimuli for rats, but other studies have also used visual cues. My colleague Helene Richter, for example, conducted what she termed 'optimism research' using automated mouse cages with built-in screens. These cages can be programmed so that a bar appears at the top or the bottom of the screen. If the bar appears at the top, the mouse must touch the screen on the left to receive a food reward. If the bar appears on the bottom, the mouse must touch the screen on the right to avoid an unpleasant noise. Once the mouse has learned these relationships, the crucial question arises: how will it behave when a bar appears in the middle of the screen? Will it associate the ambiguous stimulus with a reward or a punishment? Much like an anxious, melancholic, or depressed person, a mouse who presses the right side of the screen in an attempt to avoid the unpleasant sound likely has a negative base mood. If it reaches for the left side of the screen, on the other hand, its base mood is likely positive.

Through such studies, we can attempt to ask animals which of their experiences and housing conditions influence their world-view. Lifestyle factors that contribute to a more optimistic outlook are likely associated with positive emotions and a higher quality of life. Previous research into mammals and birds has supported this view: environmental enrichment promotes optimism in rhesus monkeys, pigs, and starlings, while isolation leads to more pessimism in dogs. Starlings who exhibit stereotypies are more pessimistic than conspecifics without behavioural disorders. Unsurprisingly, cows marked with a branding iron are also more pessimistic in the short term.

EMOTIONS

Animal emotions are one of the most current, exciting, and difficult topics in behavioural biology. By now most behavioural scientists would probably agree that animals, and vertebrates in particular, have emotions that can be scientifically studied. One major reason for this is that the limbic system – the part of the brain responsible for generating emotions in humans – is a very ancient structure that was present in our non-human ancestors and is also found in all mammals and vertebrates. The same neural pathways are activated in humans and animals to evoke basal emotions; the same chemical messages travel between neurons, and the same genes are switched on and off to regulate emotional states.

This astonishing correspondence can be seen in the two best-studied emotions: anxiety and fear. In the case of a concrete threat, for example, when a human, monkey, or rodent spots a snake or a cat, the sight triggers the same cascade of reactions: the heart begins to race, breathing begins to quicken, stress hormones are released, the facial muscles contort, and all the creature's attention turns to the present danger. The reaction is consistent across species down to the molecular level: in the region of the brain, neurons, synapses, chemical messengers, and genes, the processes are the same. With so many physical processes in common, it seems only logical to assume that humans and animals are feeling the same emotion when under threat: fear.

This assumption is also supported by the effect of certain drugs that influence the anxiety circuits in human and animal brains and lead to comparable changes in behaviour. Anxiolytics, or anti-anxiety drugs, allow people to become more courageous and comfortable with risk. When the same drugs are given to mice and rats, the rodents venture into bright, open, and unprotected areas that they would otherwise avoid. Anxiety-inducing substances, on the other hand, cause both humans and rats to retreat to protected places. The similarity that these psychotropic drugs have on brain activity and behaviour across such species suggests that these creatures experience anxiety very similarly.

But human emotions, of course, do not stop at anxiety and fear: we all know joy, sadness, disgust, anger, frustration, jealousy, shame, pride, regret, and many others. A common and highly controversial question is whether all human emotions have an equivalent in non-human mammals. As we have seen, fear and anxiety involve comparable neural circuits in both humans and animals, but how the brain produces most other emotions is poorly understood.

To still make statements about these more complex emotions, we must extrapolate from our knowledge of human behaviour. We often feel frustration, for example, when we were expecting something positive that does not occur, and we tend to behave aggressively as a result. Very similar correlations have been observed in various animal species: pigeons, rats, or squirrel monkeys, for example, can learn to press a lever to receive a food pellet whenever a bulb lights up in their enclosure. But if the animals suddenly do not receive a reward after pressing the lever, they indeed become highly aggressive towards whatever is nearby, be it the cage wall, the food bowl, or a conspecific. Frustration, just like fear, anxiety, or joy, seems to be an emotion that animals experience in much the same way that we do.

Jealousy is experienced as well: if the person you love suddenly becomes interested in another, you might try to recapture your beloved's attention. Dogs have been known to also behave in this way: In one study, owners were instructed to ignore their dogs and instead played with a realistic dummy that could bark, whine, and wag its tail at the push of a button. The dogs quickly reacted; whimpering, squeezing between their owner and the dummy, and acting aggressively towards it. They did not behave this way when their owner paid attention to a Halloween pumpkin or read aloud from a book. The results from this study indicate that our non-human relatives also experience jealousy.

Such approaches have limitations, of course. After all, the same behaviour in humans and animals is not always indicative of the same emotions, and the same emotions can be expressed in different ways across species. A chimpanzee that exhibits a facial expression similar to a smile is by no means pleased: it is frightened. Dolphins, meanwhile, appear to always be smiling no matter their mood; we can see anxiety on a wolf's face, while a bear does not have the necessary nerve and muscle control to express anything facially at all. If we try to infer an animal's emotions solely from the similarities of its behaviour to that of humans, we can easily fall prey to misconceptions, and run the risk of attributing more of our own emotions to animals that look similar to us (like primates) and are capable of facial expressions (like dogs) than to animals like bats or moles who express themselves in much more foreign ways.

There is another argument that calls for caution: emotions, like all other characteristics, have evolved through natural selection, helping animals to adapt to their environment, survive, and reproduce. We tend to understand fear and anxiety as negative, for instance, but it is easy to see why humans and animals have evolved to feel them: animals that were fearful in dangerous environments were more likely to survive and pass their genes onto the next generation than conspecifics who were not.

Thus, it is possible that emotions like fear, anxiety, and joy are universal among mammals. But different species live in very different habitats: whales in the ocean, bats in the air, polar bears in the Arctic, lions on the savannah. If emotions help a creature to better adapt to its environment, then it is quite possible that different emotions have also developed in different species across these habitats – all mammals might have a set of common emotions, but humans could also very well feel things that whales or elephants have never felt before, and likewise whales, elephants, bats, or cats might feel things that we cannot imagine.

Many behavioural scientists therefore do not consider searching for analogues between humans and animals to be entirely helpful. Such analogues cannot tell us for sure whether animals experience certain complex emotions – regret or shame, for example – in the same way that humans do. Rather, these scientists advocate focusing primarily on two aspects of emotions: valence and arousal. Is the animal in a positive or negative emotional state? Is its arousal high or low? This has become more possible with the research methods we have today, enabling us to better understand the conditions under which positive and negative emotions occur.

SPECIES-APPROPRIATE AND HUMANE LIVES

There has rarely been a public debate about animal welfare in recent years in which the concept of humane and species-appropriate living conditions has not taken centre stage. Indeed, hardly anyone would argue against behaving towards animals with these important standards in mind – but what exactly do they mean? To answer this, we must clearly define both concepts and distinguish between wild and domestic animals.

Behavioural biology defines 'species-appropriate' as behaviours that have evolved through natural selection to help animals adapt to their habitats and maximise their reproductive success. It is thus species-appropriate for marmots to hibernate, squirrels to gather food for the winter, or birds of paradise to court mates with elaborate displays. As we will see in Chapter 7, it is also species-appropriate for mating rivals to injure opponents in fights or for males of certain species to kill their predecessor's young after taking control of a pack. Ultimately, all behaviour that occurs in an animal's natural habitat and serves to maximise fitness is species-appropriate, regardless of how we would ethically evaluate it if carried out by humans.

The ways in which wild animals live in their natural habitats is considered species-appropriate. What is often overlooked in public debates, however, is that a species-appropriate life in the wild is usually racked with stress and danger. This is par for the course: in every animal species, far more individuals are usually born in a generation than are needed to carry a population into the next. Individuals are thus competing for resources and mates, and often dying in the process. In fact, numerous studies show that extreme stress reactions, injuries, disease, and shortened lifespans are not the exception in the wild, but the rule. Life in an animal's natural habitat is always species-appropriate, but it is not necessarily conducive to welfare – natural selection does not primarily favour the welfare of individuals but works to maximise their fitness over their lifetime. Accordingly, the focus of maintaining species-appropriate conditions for wildlife is on the protection and conservation of populations over individuals. The goal of such efforts is to preserve wildlife habitats so that stable and self-sustaining populations can exist without the need for human intervention.

The situation is different for animals in human care, where the focus is on the welfare of the individuals for which we are responsible. In this context, all conditions that promote such welfare can be considered humane. Thus, while the term 'species-appropriate' refers primarily to whole populations in their natural habitat, 'humane' refers to individuals in human care.

Most animals in human care are domesticated. Farm, lab, therapy, and house animals have all evolved from their wild ancestors only through a long process of domestication. Humans have bred formerly wild animals over many generations and selected for certain characteristics – be it milk yield, docility, or alertness – thus transforming the wolf into the dog, the wildcat into the house cat, the wild horse into the domestic horse, the mouflon into the sheep, and the boar into the pig.

Biologically speaking, domestic animals and their wild ancestral forms still belong to the same species; indeed, it is common for wolves and dogs or wild and domestic guinea pigs to mate and produce reproductively viable offspring. But the domestication process always results in changes in appearance, physiology, and behaviour. Typical characteristics arise, which differentiate domestic animals from their wild ancestors. The physical variability of all wolves on Earth, for example, is nothing compared to the difference in appearance between a Great Dane and a Chihuahua. The brains of domestic animals also tend to be smaller than those of their wild ancestors – in some species, by up to 30 per cent.

Domestic animals also tend to be less aggressive and more agreeable than their wild counterparts, as breeders have selected for characteristics that make them calmer and easier to handle. They are also louder and less aware of their surroundings, characteristics which would not allow an animal to survive in the wild for very long. On the physiological level, domestic animals experience far fewer stress reactions, releasing significantly less cortisol and adrenaline in comparable situations than their wild ancestors.

These changes are not deficits – rather, they are what have enabled domestic animals to survive under conditions created by humans. While wild animals are optimally adapted to their environment through natural selection, domestic animals have adapted to their environment through the domestication process. It is thus much easier to humanely keep a domesticated animal in human care than a wild one.

It is also true that domestic animals would be significantly imperilled if they had to live under the conditions of their wild counterparts. A domestic guinea pig who suddenly found itself in the natural habitat of the wild guinea pig in South America would probably have no chance of surviving there. Domesticated animals are generally so different from their wild conspecifics that their species-appropriate way of life in the wild can no longer serve as a blueprint for a humane life in captivity. Therefore, the standards for keeping domesticated animals humanely are based on a generous system of human devising rather than on the ecological niche of its wild ancestor.

CONCLUSION

Animal welfare is a central theme of modern behavioural biology, which covers not only physical but also psychological well-being. Whether an animal is suffering or thriving can be objectively and reliably determined with the help of various methods: measuring hormones can indicate an animal's stress level, while observing spontaneous behaviour can key a scientist into whether an animal is doing well or suffers from behavioural disorders. Preference tests can give scientists a glimpse into how animals see the world – what they like, dislike, and find important. Finally, methods assuming an interaction of cognition and emotion can be used to infer an animal's emotional state. Using a combination of these methods can allow a researcher to make well-founded statements about an animal's welfare, thus enabling us to create more humane animal housing conditions that promote positive experiences.

Wild animals are well adapted to their environments through natural selection and lead species-appropriate lives in their habitats. But most animals in human care are not wild, rather they have been adapted to human conditions through generations of domestication. In the process, their appearances, physiology, and behaviour have changed, and they have developed the ability to adapt well to environments created by humans. Still, this does not guarantee that they will live humanely under captivity – humans must always still create suitable conditions for them.

CHAPTER 4

Nature Versus Nurture

Genes, Environment, and Behaviour: New Answers to an Old Question

QUESTIONS OVER THE ORIGIN OF BEHAVIOUR HAVE troubled science and society for generations. For one, does behaviour arise from instinct, or learning? What role do genes play in behavioural development? How important is environment? Much research has been done on this topic, and there is even more speculation. In recent years, however, new methods from genetic engineering have revealed exciting new answers to the old riddle. But let's start at the beginning.

BEHAVIOURISTS AND CLASSICAL ETHOLOGISTS

Two distinct schools dominated the early days of behavioural biology: the European classical ethologists, including Lorenz and Tinbergen, and the North American behaviourists, the most famous of whom were Watson and Skinner. The ethologists had extensive biological training and studied a wide variety of species across the animal kingdom, from the grey goose to the stickleback to the digger wasp. They were particularly fascinated by how these animals behave in perfectly adapted ways out of pure instinct: for instance, the offspring of the digger wasp hatch in the spring when their parents have been dead since the previous summer. Female wasps will mate and then perform a whole series of complicated behaviours immediately after hatching: digging a nesting hole, building storage cells, hunting and killing prey, depositing the prey into the cells to feed their offspring, laying eggs in the cells, and finally covering them up. All this must be done in the few weeks before the wasps die. It is impossible for these creatures to have learned these behaviours from their parents: they were long since dead. It is also unlikely that they would be able to cleanly execute this sequence if they had had to learn through trial and error. Thus, the ethologists concluded that this kind of behaviour was instinctive, or innate.

Unlike the European ethologists, the behaviourists were psychologists by training, and studied animal behaviour primarily to better understand humans. They were interested above all in general laws of learning, and limited their research to a few species, in particular rats and pigeons, which could be observed in laboratory settings.

Behaviourists were fascinated by the fact that the animals they studied could learn highly complicated things if they were rewarded or punished in the right way. Two pigeons called Jack and Jill, for example, were able to 'talk' to each other using symbols after being individually trained to do so for one to three hours a day over five weeks. When the two pigeons were placed in adjacent cages after the training, Jack could be prompted to peck at a button in his cage on which the words 'What colour?' were written. Jill would then peer behind a curtain in her cage to see which of three colours – red, green, or yellow – had lit up. She would then peck at one of the three buttons on which the name of the correct colour was written in black and white. Jack, meanwhile, watched this process unfold from his cage, where he couldn't see which colour was behind the curtain, but he could see which button Jill had pressed. After he had finished observing, he would peck at a 'Thank you' button, which was clearly visible to Jill, and Jill would be rewarded with food. Jack would then peck at a red, green, or yellow-coloured button in his cage, which corresponded to the one Jill had previously chosen symbolically, and he was also rewarded with food. Thus, through the use of learned symbols, the pigeons were able to transmit information about hidden colours to each other, resulting in an ongoing 'conversation'.

Classical ethologists and behaviourists fought bitterly for decades over the question of learned versus innate behaviour. Members of both schools would take extreme positions, claiming almost all behaviour was a product of one or the other. From today's perspective we can see how the ethologists clearly overemphasised the proportion of behaviour that is innate, while the behaviourists tended to underestimate it.

An important dimension to the issue that was lost amid this polarisation is that instincts can be modified through learning. For example, adult herring gulls have a red spot on their yellow beaks, which their chicks innately start pecking at when their parents enter the nest, indicating that they want to be fed. If inexperienced chicks are shown a dummy beak made of cardboard or wood with different coloured spots, they instinctively prefer red and avoid blue. But if the response to blue is rewarded over red, the chicks' preference quickly changes.

On the other hand, an animal can only learn something for which it has an innate predisposition. Herring gulls learn to distinguish their offspring from other chicks a few days after they hatch. Their close relatives the kittiwakes, on the other hand, cannot distinguish their own offspring even after four weeks. But this is no surprise: herring gulls breed on the ground in large colonies in which the nests of different pairs are only a few inches apart. The chicks often walk around the colonies, creating a selection pressure for the adults to be able to pick out their own offspring. Kittiwakes, however, breed on smaller rocky ledges, which only have space for a single nest with two young. The chicks that the parents find on their ledge, then, are automatically their own; there is no need for the adults to be able to distinguish them. Differences between the two species' learning abilities are thus explained by their respective ways of life. Natural selection has allowed different predispositions to develop that determine the limits and possibilities of learning.

The beliefs of the ethologists and behaviourists have converged over the years, and it is now generally accepted that complex behaviour arises from the interplay between instinct and learning. Polecats, for example, instinctively know to chase, knock down, grab, and shake a rat to death, but they only learn their signature neck bite through experience. Young ducks and geese instinctively know that after hatching they must follow an object that moves and makes sounds around them. Who that object is – their mother or Konrad Lorenz – must be learned. Adult male guinea pigs instinctively know how to court and mate with a female, but which females of their group they may approach with sexual intent and which they may not is knowledge they have to acquire.

A particularly impressive example of the ways in which instinct and learning intertwine is a series of studies on the warning calls of an African monkey species conducted by American primatologists Dorothy Cheney and Robert Seyfarth. Vervet monkeys not only use their calls to warn conspecifics of predators but can also communicate the kind of enemy approaching. One type of call indicates the arrival of dangerous mammals – leopards especially – and conspecifics who hear it immediately climb the nearest tree. An eagle, on the other hand, is indicated by a different sound, to which all the members of a group react by looking up or hiding in the bushes. A third sound warns only of snakes and prompts the monkeys to search the ground. Interestingly, the animals only learn the meaning of each sound and the proper reaction to it in the course of their behavioural development. When an adult in the group calls out 'snake', 'leopard', or 'eagle', all members immediately respond accordingly. But when a baby monkey calls out 'leopard', for example, the adults do not immediately climb into the trees, but first look to the baby's mother and watch her response. The adults know that babies tend to make mistakes when they are still learning.

RELEASING MECHANISMS IN HUMANS AND ANIMALS

As we saw in the first chapter, classical ethologists developed important models for how certain behaviours are released. In their view, a key stimulus in an animal's environment activates what they termed an innate releasing mechanism, which then leads the animal to execute a certain behaviour. These mechanisms act as a lock, so to speak, on a certain behavioural response, for which the stimulus is the key. Which environmental stimuli have a triggering effect can be tested with what are called dummy experiments.

We have already seen how a dummy experiment functions in the example of male sticklebacks, who react violently to the sight of the red underbelly on a conspecific. A piece of wood with a red-painted underside proved to elicit the same level of aggression in these creatures, indicating that the stickleback does not respond to the overall appearance of a rival, but only to the red of its underbelly. Similarly, robins will not only defend their territory against a rival with red-breast plumage but also against a branch fitted with red feather tufts. The effect of this sort of key stimulus can be so far-reaching that songbirds will not only feed their offspring when they open their throats and present their colourful markings but will also feed test tubes inserted into the nest and marked with filter paper coloured in the same way.

If animals react properly upon first encountering key stimuli, such responses appear to be instinctive, encoded in the genes of a species and passed down through generations. The original response to a key stimulus can nonetheless be altered by learning, as we have seen in the young herring gulls who could change their pecking response according to the colour of a beak. Herring gulls have even been observed reversing their response to a releasing stimulus completely: when these birds first witness an underwater explosion, caused by blast fishing – a destructive fishing technique, which is fortunately now forbidden in most countries – they instinctively flee. But they quickly learn that dead or stunned fish will float to the surface of the blast site soon afterwards, and later on will accordingly flock to the sound of an explosion to take advantage of the easy prey.

It is often asked whether humans also exhibit such innate behavioural responses to key stimuli. In fact, there is much evidence that we do: for example, almost all humans across the widest diversity of cultures find babies cute. The sight of a baby elicits positive emotions in us: we want to care for it. How does this happen? Lorenz suggested that this reaction is innate, evoked by a combination of infant facial features he termed the 'baby schema': large eyes, a high forehead, a small mouth and nose, and large, chubby cheeks. Whenever a human is confronted with these features, Lorenz postulated, they will experience positive, tender, and caring emotions quasi-reflexively. But is this true?

A few years ago, my doctoral student Melanie Glocker, along with an international team of behavioural biologists and neuroscientists, conducted a remarkable study in response to this question. First, Glocker selected 17 photos of baby faces. Using special software similar to that employed by plastic surgeons, each image was then manipulated into three different versions: first, the original photo; second, a version with enhanced features from the baby schema (a rounder face, higher forehead, larger eyes, and a smaller nose and mouth), and third, a version in which these key features were attenuated (a narrower face, lower fore-head, smaller eyes, and a larger nose and mouth). The team then showed all 51 of these images to 122 college students in Philadelphia in a random order for four seconds each and asked: 'On a scale of one to five, how cute is this infant?' and, 'How much does this infant make you feel that you would like to take care of it?'

The results of the study were clear: the more exaggerated the baby schema, the cuter the students found the child. Male and female students did not differ in their assessment. Responses to the question about caring for the child yielded similar results: when presented with the original photos, male and female respondents expressed a similarly strong desire to do so. The photos with attenuated features were met with a decrease in interest in both sexes, while the exaggerated photos had the opposite effect – but only in females.

Overall, these results certainly supported what Lorenz had postulated almost 70 years prior. But how exactly does the baby schema trigger positive emotions – indeed, happiness – in us? Glocker and her team found an answer to this question as well. In a follow-up study, the researchers showed 16 women images of babies with enhanced, reduced, or unchanged features. This time, however, the subjects were in a mag-netic resonance imaging scanner, enabling the team to use a technique called functional magnetic resonance imaging to determine which regions of the brain were particularly active when the baby photos were viewed. The exciting result was that the more pronounced the baby schema, the more intensively the brain's reward centre – an area in the lower forebrain known as the nucleus accumbens – was activated. As brain researchers have long known, activity in this area triggers feelings of happiness and, interestingly, also plays a crucial role in addiction. The sight of a liquor bottle, for instance, is likely to lead to similar brain activity in alcoholics. Accordingly, the German newspapers *Die Welt* and the *Hamburger Abendblatt* reported on these studies under the headlines: 'Googly Eyes Work the Same as Narcotics' and 'Googly Eyes Prime Women's Brains for Happiness.'

The baby schema is present not only in humans but also in animals and even objects. Its effect on humans is so strong that even looking at non-human objects and creatures triggers similar emotions to looking at a baby. Indeed, the baby schema is present in many juvenile animals: tiger, lion, wolf, and fox cubs all possess it, and we perceive these offspring to be much cuter than their parents. Even the adults of some species are characterised by pronounced childlike features: deer, pandas, and koalas all possess the epitome of this combination of features. Knowing the effect of these faces, it is not surprising that the World Wildlife Fund uses the panda as its logo and not an equally endangered viper. The film industry, too, has capitalised on the power of the baby schema for decades: just think of Mickey Mouse or Nemo. Even many of our consumer products bear traces of it – the headlights of the Volkswagen Bug, for example, look just like the large eyes of a child.

In sum, these studies on the baby schema demonstrate how certain environmental stimuli can trigger predictable responses in humans, too. This seems to be true for almost all people in all cultures that have ever been studied. And the baby schema affects adults and children alike – even four-month-old babies seem to respond to it. There is thus much to suggest that these are innate responses to a key stimulus. Nonetheless, like an animal's innate response to certain stimuli, a person's reaction to the baby schema can be shaped and modified by culture and experience throughout the course of their life.

CLASSICAL METHODS OF STUDYING INNATE BEHAVIOUR

The examples in this chapter so far have demonstrated how scientists can indirectly conclude whether a certain behaviour is innate or not. Simply put, if an animal perfectly executes a behaviour that it had no prior opportunity to learn, and if that same behaviour occurs in the same way among all members of its species, then it is most likely innate: take, for example, the elaborate webs that many spiders are able to weave from the first try. While such indirect conclusions are key to understanding certain behaviours, behavioural geneticists take a different approach. As the title suggests, their field of research looks specifically at how an individual's genes influence its behaviour. The findings and methods of this discipline will be discussed in the following section.

At the outset, it is important to clarify that no behaviour is either purely genetic or purely a product of environment. As previously mentioned, all behaviour is a result of the interplay between both forces. Nonetheless, genetics or environment alone can also cause differences in the behaviour of two individuals. If a zebra finch raised by its parents is given the choice between mating with another zebra finch or a Bengalese finch, it will choose its conspecific. But if it is raised by Bengalese finches and then given the same choice, it will prefer the Bengalese finch. Does the fact that a zebra finch chooses a mate have anything to do with its genes? Of course! The ability to choose requires a brain that could never have evolved without the right information encoded in its genome. Every behaviour, no matter how simple, ultimately arises from the activity of nerve and muscle cells which are, in turn, controlled by genes. But does the difference in the birds' choice of mate have anything to do with genetics? Not at all – these different choices are a product of the parents that raised the birds, a purely environmental factor. The crucial question of behavioural genetics is not whether a certain behaviour is a product of genetics, but whether animals and humans who differ in their genetic constitution therefore also differ in their behaviour.

How can we test whether genes are causally involved in such behavioural differences? A few decades ago, crossbreeding experiments yielded the most promising results (although they no longer play a significant role in current research). In principle, crossbreeding entails mating animals of two closely related species and comparing the parents' behaviour to that of the offspring's. Male pheasants, for example, bolt upright and point their head and tail feathers skyward when they crow, while male domestic roosters lean diagonally and point their bill and tail feathers towards the ground. When these species are crossbred, their offspring assume a crowing posture that is exactly between the two parent species.

The offspring that result from crossbreeding are in most cases incapable of reproduction. If, in rare cases, they are still able to breed, we can draw detailed conclusions about the original species' genomes from the results. A well-known example is the crossbreeding of two closely related cricket species; one of which is highly aggressive, the other relatively calm. All the offspring from this first round of crossing turn out to be aggressive. But if these offspring are then bred with each other, about three-quarters of their offspring are aggressive and the rest calm. These results suggest that only one of the several thousand genes in a cricket's genome determines whether the animal is aggressive or docile. How can we conclude this?

Every gene consists of two alleles. In the aggressive species of cricket, both alleles of the gene that determine the animals' temperaments are encoded for aggressive behaviour. In the peaceful species, both alleles are encoded for docility. When the species are crossbred, the offspring receive one allele from each parent, and thus possess one aggressive and one peaceful allele. Since all the offspring of the first generation act aggressively, we can conclude that the aggressive allele is the dominant one. Thus, when the crickets of this new generation are bred, four genetically distinct groups can result: the first receives an aggressive allele from its father and mother, the second an aggressive one from its father and a peaceful one from its mother, the third a peaceful one from its father and an aggressive one from its mother, and the fourth a peaceful allele from both. Because the aggressive alleles are dominant, any group that has one or two aggressive alleles will behave aggressively. Those that receive two peaceful alleles, however, will behave calmly with one another.

Another traditional way of gaining insight into the hereditary basis of behavioural differences is selective breeding. This process consists of choosing animals with certain behavioural traits to mate with one another, which, if the traits have a heritable component, will become more and more pronounced over generations. For example, one study tested the ability of each rat in a population to find its way through a maze. As might be expected, several rats were very intelligent, several were quite unintelligent, and many were somewhere in the middle. The most intelligent males were then selected to mate with the most intelligent females, and their offspring were tested in the maze. The most intelligent rats of this second generation were subsequently selected and bred. The same process was seen through in parallel with the least intelligent rats. The results of the two selection processes were almost frightening: after only seven generations, two clear populations had emerged. The effect was so strong that even the least intelligent of the genetically 'clever' rats was smarter than the most intelligent of the genetically 'non-clever' rats. Studies on numerous other animal species have also confirmed how effective artificial selection can be in exaggerating certain behavioural traits. This method can also be used to breed docile and aggressive mice, tame and wild mink, and even loud and quiet crickets within only a few generations.

We have already discussed one particular experiment in selective breeding, which has gone on for several thousand years: domestication. For centuries, wild animals have been bred for certain desired characteristics, quickly leading to significant changes in appearance, physiology, and behaviour.

MODERN BEHAVIOURAL GENETICS

Behavioural genetics research has changed drastically since the days of crossbreeding experiments. The entire human genome and those of many animal species have since been completely decoded and, although in principle all individuals of a species have the same number of genes, the structure of these genes can differ significantly. Many research teams around the world are now trying to understand whether certain genetic differences actually lead to discrepancies in behaviour and, if so, what path they take to causing these changes. What does this line of research look like in concrete terms?

About 25 years ago, an article published in *Science* sent shockwaves through the behavioural research world. In it, a team of Dutch and American scientists reported on five distantly related Dutch men who lived in different parts of the country, all of whom possessed a certain level of intellectual disability and conspicuous impulsivity and aggression in response to anger, fear, or frustration.

Psychiatrists have long known that the malfunction of neurotransmitters like serotonin and noradrenaline can trigger a strikingly high level of aggression in humans. How do such malfunctions occur? A protein called monoamine oxidase A, or MAOA for short, exists to properly degrade these chemicals after they have effectively passed on information from one nerve cell to another. At the time of the article, the gene that carries the information for the production of MAOA had already been discovered, and the researchers suspected that it was defective in the five Dutch men they had identified. The investigation supported their theory: all the men carried a tiny error in this gene, meaning that their cells did not produce MAOA.

The study's sample size of five was admittedly too small for the researchers to draw any general conclusions. It had also not been possible to directly measure the way these men's brains metabolised serotonin and noradrenaline, so any statement the researchers made in this regard was based in speculation rather than fact. (Even today, more than a quarter of a century later, it is still not clear how the brain uses these neurotransmitters to control aggressive behaviour.) The study nevertheless clearly laid out how genes may influence behaviour, carrying the information for the production of certain proteins that are involved in the brain's process of behavioural control. Even tiny alterations like a point mutation in a single gene can lead to serious changes in behaviour.

Two years later, *Science* published another study that built upon the logic of the first. This time, a team of American, French, and Swiss scientists wanted to know whether eliminating the MAOA gene actually leads to the predicted changes in the concentrations of serotonin and noradrenaline that in turn increase an individual's levels of aggression. Ethical concerns prevent such a study from being conducted on humans, so the research team used mice as a model, precisely modifying their genetic material so that the gene that guided the production of MAOA no longer functioned. (The several thousand other genes in the animals' sequence remained unaffected.) As expected, these mice no longer produced MAOA, experienced drastically higher concentrations of serotonin and noradrenaline in their brains, and demonstrated significantly increased aggression. Groups of mice with the defective MAOA gene often got into fights, while their conspecifics whose gene was intact generally settled conflicts peacefully. When mice with the defective gene encountered an unfamiliar conspecific, they immediately attacked; their unmodified counterparts usually behaved with much more reserve. Altogether, this study confirmed that a change in just one gene of thousands can lead to significant behavioural changes.

Over the past 25 years it has become increasingly clear that genes can influence almost every kind of behaviour. We now know that certain genes mediate whether animals tend to wake up early or sleep in, for example; others determine how quickly they can learn a task, and others still how much energy they put into caring for their offspring. Certain genes modulate sexual behaviour; others influence how friendly animals are with conspecifics.

But how comparable are genetic studies of humans and animals? One particularly astonishing similarity is in the role of genes in generating emotions. About 25 years ago, the neuroscientist and psychiatrist Klaus-Peter Lesch and his colleagues in Würzburg described the variants of the gene responsible for the creation of a serotonin transporter (SERT) in humans. SERT is a protein, which transports released serotonin back into a nerve cell where it is then stored for further release. The researchers found that individuals can carry two long alleles on this gene, two short ones, or one short and one long one. The short allele is associated with a reduced production of SERT. The differences in the structure of the SERT gene had clear effects on mood: carriers of short SERT alleles were significantly more anxious than carriers of long ones.

Interestingly, very similar variants of the SERT gene are also found in rhesus monkeys. Like in humans, the carriers of the short alleles are markedly more fearful than carriers of the long ones.

To better understand how the SERT gene can have such a strong influence on behaviour, Lesch's team designed a 'SERT knockout mouse'. As the name implies, the SERT gene in these mice was switched off in a genetics lab. Breeding two SERT-modified mice will result in SERT-deficient offspring. A SERT-modified mouse and an unmodified conspecific, on the other hand, will produce offspring that possess one intact and one defective allele on the SERT gene. Through clever breeding, three different types of offspring can be produced: mice with zero, one, or two functioning alleles, which all produce corresponding levels of SERT in the brain.

Detailed examinations initially showed that mice of all three genotypes were healthy, had developed normally, and possessed perfectly functioning senses. But further investigation revealed clear differences between them: mice with two intact SERT alleles were generally less anxious than conspecifics with two defective ones, exploring unfamiliar territory more boldly, venturing more often into open areas, and learning more quickly which kinds of situations were dangerous. Animals with two defective alleles correspondingly avoided bright and unprotected areas and took much longer to erase negative experiences from their memories, while mice with only one functioning SERT allele generally behaved somewhere between the other two genotypes. The effects of the SERT gene on mood and behaviour thus show striking similarities between humans, monkeys, and mice.

The studies on SERT knockout mice also confirmed another important aspect of the connection between genes and behaviour that applies equally to humans and animals: a single gene usually affects not just a single behavioural trait, but a wide variety of them. Changes in the SERT gene affect how anxious or curious an animal is, how boldly it ventures into new situations, how aggressively it approaches conspecifics, how stressed it becomes in response to environmental changes, how quickly it learns to cope with change, and how optimistically or pessimistically it assesses ambiguous situations.

DO GENES DETERMINE BEHAVIOUR?

If changes to a single gene can result in such drastic changes in behaviour, the question follows: is behaviour determined by genes? To put it another way: is an animal's behaviour ultimately decided by the genetic material it receives from its parents? Before answering this controversial question, it is important to address the fact that evidence of such single-gene transformations comes from studies in which all other factors are held equal. The effect of the defective SERT gene was determined by comparing two groups of mice; one possessing a defective gene, the other a functioning one. But these mice did not differ in the rest of their characteristics – their thousands of other genes were the same; they were the same sex, the same age, and lived under the same conditions. They ate the same food, lived at a constant temperature, and all relied on the lights turning on at 08:00 every morning and turning off again at 20:00 at night. When these animals were tested for their behaviour, it was therefore sensible to conclude that the genetic change accounted for their difference in behaviour.

We already know what occurs when an animal's environment changes while its genes are kept constant – we addressed this in the last chapter, which dealt with how shifts in environment affect behaviour and welfare. If genetically identical mice of the same sex and age live in either super-enriched or standard housing conditions, the differences in behaviour are like night and day: in the enriched environment the animals were friendly with one another, playing frequently and rarely acting with aggression. They were brave, learned quickly, and had positive life experiences. Mice in the standard environment, on the other hand, were clearly experiencing the opposite. Thus, identical genes did not lead to identical behaviour at all: the environment had the decisive influence on how the animals behaved.

Thus, we ask again: do genes determine behaviour? The answer is ultimately no. Like the environment, genetic predisposition can influence behaviour, but cannot determine it alone. Behaviour always results from the interaction between the two, and can indeed be more accurately mapped – and this is an innovation of the last few years – by tracing predisposition to the level of individual genes.

THE INTERPLAY OF GENES AND ENVIRONMENT: INTELLIGENCE IN RATS

This core insight into the genesis of behaviour is not a new one. As early as 1958, a fantastic and oft-forgotten study on the subject was published in the *Canadian Journal of Psychology*. In it, researchers looked at two different lines of rats that had been bred over generations according to their learning ability. Consequently, the animals from one line were genetically 'intelligent' and able to solve a maze with relatively few mistakes, while the animals from the other line were genetically 'unintelligent' and often lost their way. Huge differences in the learning performance of the two lines were thus evident – but only if all the subjects had been raised in a stimulating environment. If the animals had lived in barren, low-stimulus enclosures, the two lines no longer showed any difference: while the learning performance of the 'intelligent' rats deteriorated dramatically under these conditions, the genetically 'unintelligent' ones showed no real further impairment. In contrast, an enriched environment had a significantly stronger effect on the 'unintelligent' rats when compared with the 'intelligent' ones – the learning performance of the latter hardly improved, but the former proved to be quite intelligent, making only marginally more errors than the other group. When the researchers compared 'intelligent' rats who had been raised in a low-stimulus environment to 'unintelligent' ones from a rich one, the latter group proved to be even more clever than the former.

This study demonstrates that a genetic predisposition for certain learning abilities certainly exists. How intelligent an animal ultimately is, however, is a result of the interaction of such predispositions with the environment. Genes alone do not determine the intelligence of a rat.

THE INTERPLAY OF GENES AND ENVIRONMENT: WHAT MICE CAN TEACH US ABOUT ALZHEIMER'S DISEASE

Half a century after the study in the *Canadian Journal of Psychology*, a body of research on Alzheimer's disease emphatically confirmed the role that environment plays in determining the extent to which a genetic predisposition is expressed. Although Alzheimer's disease is not primarily genetic in the majority of cases, there is a rare form of the condition known as familial Alzheimer's disease, which is caused by mutations on the APP gene. If a person has such errors in their genome, malignant proteins will deposit in the brain at an early age, causing symptoms of Alzheimer's disease with almost 100 per cent certainty.

Mice do not normally possess this defect; they do not normally develop Alzheimer's disease. But when Canadian scientists introduced the defective human gene into the genome of lab mice, malignant proteins began to deposit in the animals' brains, which were no different from the ones present in human patients. The cognitive performance of these mice revealed another parallel: the mice showed significantly poorer orientation and memory ability. As in humans, these symptoms did not appear until adulthood; the mice developed normally and health-ily through adolescence. Later studies revealed additional features to the mice's behaviour that are also common in Alzheimer's disease patients, such as an altered sleep cycle, hyperactivity, and startling movement stereotypies; their levels of certain stress hormones were also significantly elevated. Altogether, these studies demonstrated how alterations to a single gene can cause mice to experience symptoms on a similar time-scale to humans.

My team had been studying the effects of super-enriched housing on the behaviour and welfare of mice when these results were published. We took great interest in them, and assembled an interdisciplinary group of brain scientists, physicians, and behavioural biologists to answer the question: Could an enriched environment also have positive effects on Alzheimer's mice? And might it be possible that the development and progression of Alzheimer's disease symptoms have a dynamic relationship to genes and environment instead of being caused by a defective APP gene alone?

Our Canadian colleagues kindly provided us with some of their Alzheimer's mice, from which we were able to build a colony through targeting mating. To investigate the effect of environmental enrichment on these mice, we compared two different groups, one of which lived in standard housing, the other in super-enriched enclosures, which gave them access to a playroom for several hours a day. These playrooms were equipped with a running wheel, a ladder, twisted tubes, balls, and scarves, which were rotated in and out on a daily cycle. The mice occupied themselves with these playthings intensely and seemed to enjoy the environment.

The enriched environment certainly had exceedingly positive effects on the brains of the mice. As the medical researchers on the team were able to show, the mice from the stimulating environment developed significantly fewer malignant deposits than their conspecifics from the standard one. The brains of the enriched mice produced significantly more new neurons (which are thought to counteract the effects of the disease) and developed a wide variety of mechanisms to activate and protect them. The enriched mice were also demonstratively more curious, exploring unknown territory much more readily than their counterparts, and showed evidence of possessing a better reasoning ability, a finding which has since been confirmed in studies by other research groups.

Thus, the same conclusion is evident once again: a genetic predisposition towards certain traits is one thing, but their expression is quite another. All the animals in our study possessed the defective APP gene, but the dynamic between their brains and environment ultimately determined how their disease developed. A varied, stimulating environment could not prevent the symptoms of Alzheimer's disease, but it was highly effective in buffering them.

But our study did not stop there. In a next step, we asked ourselves what would happen if we not only provided the Alzheimer's mice with super-enriched cages, but also built them an environment that was highly similar to where they would live in the wild – would such a spacious, nearly 'natural' habitat perhaps result in no symptoms at all?

My collaborator Lars Lewejohann set out to answer this question in a series of investigations. First, he built a large enclosure of about 3 m wide and 2 m tall, with five levels connected by ladders and ropes inside. On all levels were numerous tubes, nesting boxes, bricks, paper towels, and other objects. Frequently replenished food and water bottles were scattered throughout the room. The new enclosure contained about 30 mice of both sexes, all of whom had lived there since birth, and 40 per cent of whom possessed the defective APP gene. The rest of the animals were genetically identical apart from the mutation. Experienced observers who did not know which mice possessed which genes were then tasked with watching the animals behave for about 450 hours.

Examinations of their brains yielded disappointing results: the new environment in no way impeded the development of deposits, and in fact even more deposits formed in these mice than in their conspecifics from the standard enclosures. But the behavioural evaluation was sensational: the mice with the defective APP gene, which according to the brain scans had been causing severe Alzheimer's disease, behaved nearly indistinguishably from their unafflicted conspecifics. Adults showed no difference in their food intake, personal hygiene, or nesting habits. Their social behaviour and aggression levels hardly differed. Both groups were equally interested in their environment, and, remarkably, none of the mice showed any stereotypies or other disorders. Individuals from the Alzheimer's disease group even rose to the highest positions in their group hierarchies and were able to successfully defend their territory. Subsequent studies that used an automated system to analyse the behaviour of all the animals in the enclosure – day and night – also confirmed these findings. The Alzheimer's mice under these conditions also did not have higher levels of stress hormones than their counterparts.

In sum, Alzheimer's mice living in a nearly natural environment exhibited no behavioural impairment, despite their genetic predisposition to the degenerative disease and the numerous protein deposits accumulating in their brains. We can't with certainty say why this is: perhaps the active lifestyle in a stimulating environment could have led to the mass formation of new nerve cells, which were then used as a sort of 'cognitive reserve' to counteract the effect of the malignant deposits. What is certain is that these results once again demonstrate how behaviour is a result of a dynamic interaction, not only of genetics.

THE INTERPLAY OF GENES AND ENVIRONMENT: CLUES FROM THE SEROTONIN TRANSPORTER GENE

One of the best examples of the connection between genes, environment, and behaviour is demonstrated by research that has been done on the SERT gene. We have already discussed how SERT is a protein that brings released serotonin back into a cell, and thus plays a crucial role in deciding the effect that serotonin has on the brain. How much SERT is present in an individual is largely determined by the nature of the SERT gene: if the gene carries the information to produce just a small amount, then the humans, monkey, and mice in possession of it are more anxious and likely to develop depression than others who produce a large amount.

But even this gene has no determinant effect on behaviour, as a ground-breaking study conducted by Avshalom Caspi and his team in 2003 was able to demonstrate. In it, the researchers asked about 1000 people aged 26 whether they had experienced any seriously stressful events over the past five years. If so, how many? Had they experienced a depressive episode in the past year? Had they been diagnosed with depression? Had they engaged in suicidal ideation? Finally, all participants had their deoxyribonucleic acid (DNA) analysed to determine whether they carried two short SERT alleles, two long alleles, or one short and one long one. The results were highly interesting.

Initially, it came as no surprise that people who had experienced no severely stressful events in the past five years also reported a low level of depression, regardless of the makeup of their SERT gene. It was also unsurprising that mental-health problems increased in all participants who had recently undergone highly stressful experiences. The extent to which their mental health deteriorated, however, depended significantly on the SERT gene: Individuals who had experienced four or more periods of severe stress and carried two short SERT alleles were about twice as likely to report symptoms of depression as those with two long alleles. They were also twice as likely to have been diagnosed with depression, and three times as likely to have contemplated or attempted suicide. How did those with one short and one long allele fare? As expected, they were about halfway between the other two groups. This study is yet another poignant demonstration of the interplay we have so far discussed.

Studies on mice have also confirmed these correlations. For example, Rebecca Heiming, a researcher from our team, conducted a series of studies in which she compared the offspring of mothers who had lived in either a dangerous or a safe environment during gestation and lactation. To simulate a dangerous environment, she repeatedly exposed pregnant and lactating females to olfactory cues of unfamiliar males by introducing small amounts of their soiled bedding into the females' enclosure. Studies in behavioural ecology have shown that foreign males pose a serious threat to newborn mice, frequently killing them. Accordingly, Heiming simulated the safe environment by regularly introducing neutral bedding. The mice she used for the study had been selectively born from parents who each possessed one intact and one defective SERT allele, and thus could be one of three different genotypes in the litter, carrying one, two, or zero intact SERT alleles.

Analysis of the experiment showed the levels of anxiety and courage in the offspring to be significantly influenced by their mothers' experiences. Mothers from the dangerous environment produced more fearful and less exploratory offspring than mothers from a safe one. But genotype also played a crucial role: offspring with two defective SERT alleles were more anxious than those with one or two intact. The offspring's particular combination of SERT alleles also influenced the extent to which they were affected by their mothers' environment: those who did not possess any functioning alleles were the worse off. Similar to what we saw in Caspi's study on humans, the mice's reactions to different situations depended on their SERT gene. Once again, we see how the interaction of genetics and experiences shapes emotions and behaviour.

At first glance, a gene defect that leads to an underproduction of SERT seems to be a disadvantage. Indeed, the short SERT allele in humans is often referred to as a risk factor for anxiety disorders, as carriers of this allele are more likely to develop anxiety and depression than carriers of two long alleles. But humans are not the only creatures who can have different configurations of the SERT gene – monkeys can have combinations of a short and long allele too and exhibit similar associated behaviours to humans. From an evolutionary perspective, natural selection should gradually remove purely negative traits from a population. But this is by no means the case with SERT alleles, so possessing a short allele must bear some advantage. The American psychologist Jay Belsky has thus posed the intriguing question: could two short SERT alleles not only cause an individual to be particularly susceptible to negative events but also to positive ones? If so, carriers of the short allele would be at a disadvantage in a stressful environment but thrive in a pleasant one.

There has lately been increasing evidence that this is the case. In the Caspi study, for example, carriers of two short alleles were most at risk of psychological issues if they had experienced a great deal of stress, but they also had the fewest problems when their lives carried on without significant turmoil. Similarly, people with two short SERT alleles are more neurotic than people with two long alleles when both have had a series of negative experiences, but positive life events have the exact opposite effect: carriers of short alleles are thereafter less neurotic than carriers of long ones.

These results support a sensible suggestion that Belsky proposed a few years ago: short SERT alleles should no longer be considered as 'vulnerability genes' or 'risk alleles' but might be more appropriately conceptualised as 'plasticity genes'. Apparently the SERT gene quite generally influences how humans and animals are affected by their environments.

EPIGENETICS

Until just a few years ago, behavioural scientists clung to the central dogma that genes influence behaviour, but behaviour does not influence genes. What does this mean? In principle, offspring inherit one allele of each gene from each of their parents. As we have seen, when the environment remains constant, even slight differences in these alleles can determine the offspring's levels of intelligence, aggression, and anxiety, and thus we can see a clear path from genetics to behaviour. But individuals have very different life experiences regardless of their genes – they can learn a great deal in a stimulating environment or hardly anything in a dull one; they can be frequently confronted with aggression or live relatively free of it; they can have positive experiences that encourage them to be courageous, or have negative ones that raise their levels of anxiety and caution.

Scientists long assumed that such experiences do not lead to genetic changes. In other words, a child with two short SERT alleles has a predisposition for anxiety, and while the child may become very brave through experience, experience does not transform the short SERT alleles into long ones. When the child grows up and reproduces, it will still pass on its anxious predisposition to its kin. Indeed, the path from genes to behaviour was conceived as a one-way street – until a groundbreaking study by biologist Michael Meaney was published.

Meaney and his team at the University of Montreal studied the mothering behaviour of rats. Over time, they had noticed that 'good' and 'bad' mothers tended to exist among the animals: good mothers cared intensively for their young, licking them clean frequently, while 'bad' mothers tended to their offspring about half as much. Interestingly, these were stable character traits: those who were good or bad mothers sometimes were good or bad mothers always. The different degrees of care had clear effects on the offspring: offspring of good mothers learned better, excreted less stress hormone, and were more courageous later in life than the offspring of bad ones. That the maternal behaviour was the decisive factor in these studies was proven through what are known as exchange experiments, in which the offspring of bad mothers were given to good ones to be raised. These infants did not at all differ in behaviour from the biological offspring of the mother who raised them; the same was true when the exchange was put in place the other way. How can maternal behaviour have such drastic effects on the temperament of offspring? To understand the sensational answer that Meaney's team revealed, it is important to know a few basics of molecular genetics.

As is generally known, DNA is the carrier of genetic information. DNA consists of two parallel strands that wind around each other in a formation known as a double helix. Individual genes can be thought of as different sections of DNA that consist of four basic building blocks arranged in a row. These building blocks are called nucleotides, and each contains one of the four base amino acids: adenine, cytosine, guanine, or thymine (A, C, G, and T). In each gene, a long sequence of these four letters carries all the information for the production of substances that are crucial to the structure, maintenance, and function of an organism. Each gene contains a different combination of these nucleotides and thus codes for a different substance. These substances can be enzymes that catalyse important bodily processes, for instance, or structural proteins that give cells their shape and strength, or antibodies that fight invading pathogens. The 'instructions' for forming substances that influence behaviour are also encoded in genes, as we have seen with MAOA and SERT; other examples include genes that carry the information for the production of hormones or their receptors in the brain, which are docking sites for hormones and are thus also influential in the execution of behaviour.

Meaney and his team were able to show that the maternal behaviour of mice changes the structure of certain genes responsible for behaviour and stress responses in the brains of their offspring. While the sequence of base nucleotides in these genes remained the same, small chemical appendages known as methyl groups were attached according to the mother's behaviour. These methyl groups act like 'off' switches, deregulating a certain gene's activity. A particularly affected gene was one that carried the information for the production of an important hormone receptor. If the mother took poor care of her offspring, a switch formed on precisely this gene, partially inactivating it, and triggering a series of responses in the brain that caused the offspring to become permanently more anxious and stressed.

The mothers' actions also had a major influence on how their daughters treated their own offspring: as might be expected, daughters of good mothers were good mothers themselves, and daughters of bad mothers, bad ones. The gene in question was then switched off for this next generation, and the offspring of bad mothers were again more anxious and stressed. In sum, this study's almost unbelievable finding was that an animal's experiences can lead to fine changes in the structure of its offspring's DNA, thus causing a significant change in behaviour. These kinds of changes to the fine structure of genes that do not modify the sequence of base pairs are known as 'epigenetic'. The transmission of these changes from one generation to another is called 'non-genomic' or epigenetic inheritance.

The American scientists Brian Dias and Kerry Ressler recently conducted a sensational study, which presented further evidence of epigenetic inheritance in mice. First, the research team conditioned male mice to avoid a particular scent. They then mated these males with females who had never been exposed to the scent, and the offspring they bore never had contact with their fathers. When the children grew up, the researchers tested their reactions to certain scents, and, incredibly, the mice were hypersensitive to the same smell that their fathers had learned to avoid. Like their mothers, these mice had never been exposed to that particular smell before; nonetheless, even their offspring of the next generation had the same sensitivity to it.

How can all this be explained? Interestingly, the negative experience associated with the scent modified a gene in the sperm of the grandfathers that coded for the formation of a receptor responsible for the perception of that particular smell. This was an epigenetic change, similar to the one Meaney's team had discovered, which was passed down through generations.

More and more studies have recently been documenting this variety of genetic change, adding to a growing body of evidence that epigenetic inheritance plays a significant role in the preservation of behavioural traits across generations. The old dogma about genes and behaviour seems to have been disproven: we now know that behaviour can indeed influence genes. The exciting question of the extent of its influence, however, remains to be seen.

CONCLUSION

Behavioural biology has come a long way from the original debates over the acquisition of behaviour: while ethologists and behaviourists were once arguing over whether traits were acquired innately or through learning, scientists today are elucidating the precise ways in which genetics interacts with the environment to produce and control behaviour. One thing is clear: behaviour can be caused by both genes and the environment, and, as a rule, arises from a dynamic between the two. Indeed, the way such dynamics emerge can now be traced to the level of individual genes, and epigenetic inheritance can even result in the genetic transmission of behavioural traits across generations. We are only beginning to understand this highly complex interplay: deciphering it is undoubtedly one of the most exciting frontiers in behavioural biology.

CHAPTER 5

Clever Dogs and Ingenious Ravens

All Animals Can Learn, Many Can Think, and Some Can Even Recognise Themselves in a Mirror

A S WE SAW IN THE LAST CHAPTER, INSTINCTS AND LEARNING interact in complex ways to produce the characteristic behaviours of each species and individual. So far, we have discussed learning in a broad sense: ducklings learn who to follow after they hatch; zebra finches learn what traits their future mates should have; guinea pigs learn to get along with members of their own species; and vervet monkeys learn which warning call to sound for a leopard and which to sound for an eagle. The pigeons Jack and Jill even learned (with help) to transmit information with symbols so efficiently that it appeared as though they were having a conversation. In this chapter we will take a closer look at which cognitive capacities animals possess: whether they can not only learn but also think, and whether they, like humans, can develop self-awareness.

RICO THE BRILLIANT BORDER COLLIE

In 1999, one million viewers were made aware of the astounding learning and memory capacities of animals when they were introduced to a dog named Rico on the German television programme 'Wetten, dass ... ?' ('Wanna Bet?'). Rico, a five-year-old border collie belonging to a family by the name of Baus, was presented with 77 different toys and tasked with finding the one that the moderator had chosen. When Susanne Baus asked, 'Rico, where's the snowman? Find it! Find it!' Rico set off and examined the objects one after another. When he found the snowman, he took it in his muzzle and brought it to his owner. In the second, third, and fourth round, Rico found the Pokémon, the 'Schalke' (a small ball in the colours of the famous German football club by the same name), and the 'BVB', a similar ball in the colours of Borussia Dortmund, with the same assuredness. Rico had clearly learned to assign words to objects and to call them to mind when prompted. According to the Baus family, their dog knew the names of over 200 toys and balls and was able to fetch them on command.

Still, the history of behavioural research urges caution when it comes to assessing such cognitive performances. As we discussed in the introduction, a hundred or so years ago Wilhelm von Osten was firmly convinced that his horse, Clever Hans, could solve simple mathematical problems. But an investigation showed that, when no person who knew the answer was present, Clever Hans could not solve the problem either. With the 'Clever Hans effect' in mind, Julia Fischer and her team at the Max Planck Institute for Evolutionary Anthropology in Leipzig set out to test the Baus' border collie: was Rico really capable of such astounding feats of memory, or was he relying on unconscious cues from the people present?

In the first experiment, the research team divided the 200 toys that Rico knew into 20 groups of 10 objects each and distributed the items from the first group into a room while Susanne Baus waited with Rico next door. They then asked Susanne to instruct her dog to fetch two randomly selected toys one at a time from the experimental room so that, while Rico searched, there was no one with him who knew what he was looking for. The team gradually repeated the experiment with the other 19 groups of toys. In total, Rico was given 20 chances to retrieve the right toy 40 times from a total of 200 objects. The ingenious border collie managed it 37 times. When Clever Hans was put under scientific scrutiny, it was clear that a horse cannot do mathematics. Rico, on the other hand, was proven capable of assigning the correct names to 200 objects – an amazing achievement. But not a unique one in the animal kingdom. Great apes, dolphins, sea lions and parrots can also develop a similarly extensive 'vocabulary' if they are rigorously trained to assign words to objects. Rigour is crucial: the Baus family reportedly practised with Rico for four to six hours a day. Such training can yield increasingly astonishing outcomes. Rico's border collie colleague Betsy, for example, possesses a record-breaking vocabulary of 340 words.

Rico's achievements were by no means the result of simple conditioning, as the Leipzig team were able to demonstrate in a second experiment. Rather, the border collie used an ingenious learning method, which has long been believed to belong to human cognition alone: 'fast-mapping'. With this method, toddlers from the age of 24 months can learn an average of 10 new words a day.

What did this second experiment look like? Eight objects were distributed in an adjacent room, seven of which Rico was familiar with. The eighth object was one he had never seen before and did not know the name of. In the first round, Rico's owner asked him to retrieve a familiar object, which he, as always, accomplished effortlessly. In the second or third round she asked him to fetch an object whose name was entirely unknown to him: 'Rico! Where's the so-and-so?' Rico would then have to run into the next room, look at all eight of the objects, choose the unknown one and bring it to his owner. Rico managed to bring the right object in 7 out of 10 rounds in total. Clearly, he was able to match the new word with the unknown object through process of elimination: the object that he didn't know must have been the one that his owner had asked for.

An exciting question followed: Could Rico remember the link between new words and objects that he had only heard or seen once before? The startling answer was yes – although not perfectly. Four weeks after the experiment, an object that Rico had previously identified as unknown was placed in a room with four known and four unknown others. Rico's owner then asked him to fetch the object that he had not encountered in the last month. Nevertheless, in half the attempts he was able to retrieve it correctly. He had learned through fast-mapping to assign new terms to unknown objects, store them in his memory, and employ them four weeks later with surprising accuracy. His success rate in these experiments was right in the range that developmental psychologists have determined for three-year-old children.

No doubt Rico was capable of highly complex learning processes. But what exactly does it mean in behavioural research when an animal 'learns'? Generally speaking, learning is seen as the ability to change one's behaviour based on personal experience. Learning allows animals to adapt their behaviour to their environment, so it is unsurprising that it is widespread in the animal kingdom and can even be found in simple invertebrates, like roundworms and paramecia. Learning is always closely linked to memory, as an experience can only lead to a change in behaviour if it can be saved and recalled. We can, however, distinguish between forms of learning with varying complexity. Matching new terms with unknown objects through fast-mapping, as we saw from Rico, is a highly complicated learning process that is only likely to be found in a few species with highly developed brains. In contrast, most animals can quickly be taught, for example, that feeding time comes after the sound of a whistle.

HABITUATION: A SIMPLE FORM OF LEARNING

The simplest known form of learning is habituation. Habituation does not, strictly speaking, mean that a new behavioural response is learned, but rather that an existing one is lost. If an animal encounters a stimulus that has no positive or negative consequences, then the animal's reaction to the stimulus gradually becomes weaker. For example, if someone knocks on a pane of glass over which a snail is crawling, the snail immediately retreats into its shell. After a while it comes out again and continues on its way. If the window is tapped again, the snail will once again withdraw, although this time it will take less time to re-emerge. If one continues knocking on the glass, the time the snail takes to come out again will grow shorter and shorter until the snail no longer retreats at all: it has got used to the knocking, and learned that it has no consequences. Chaffinches reacted in the same way during a study in which they were shown a live owl for twenty minutes a day. At first sight they sounded warning calls to their conspecifics. After the so-called 'enemy' showed no reaction, the frequency of the calls decreased each day. After 10 days the finches paid almost no attention to the owl; they had learned that it was not a biologically relevant stimulus.

Such habituation responses are widespread in the animal kingdom. Although they seem unremarkable at first, they represent great advantage: they help animals avoid unnecessary behaviour, allowing them to save energy and concentrate on more essential aspects of survival. This learning process also explains, for instance, why scarecrows tend to only work for a short while.

ASSOCIATIVE LEARNING: CLASSICAL CONDITIONING

We normally connect the term 'learning' with the acquisition of new behavioural responses. This applies to the learning form that has by far been the best studied in humans and animals: associative learning, which is generally understood as the association between a previously insignificant stimulus and a punishment or reward. Through this process, an animal learns which stimuli are important and which behaviours are linked to which consequences.

Perhaps the most well-known form of associative learning is classical conditioning, which is inextricably linked with the name of the Russian scientist Ivan Petrovich Pavlov. Pavlov was a doctor who studied the digestive glands. While examining dogs for his research, he noticed that they not only secreted saliva when they were fed but also when they heard steps approaching their kennel. This observation inspired him to conduct his famous experiments, which proceeded more or less as follows: in the first step, a dog was given a portion of food, and the amount of saliva it secreted was measured. In the second step a bell sounded in the kennel, but the dog was given no food. As expected, the sound did not trigger any salivation. If the food was presented at the sound of the bell in a third step, then the dog salivated as it did the first time. If the food and the bell were repeatedly presented together, then in a fourth step the sound alone would trigger salivation. The dog had learned to respond to a previously neutral stimulus – the bell tone – by salivating. The neutral stimulus had thus become a conditioned stimulus. Accordingly, the food is termed an unconditioned stimulus.

The essential characteristic of classical conditioning is the formation of the association between the reward and the conditioned stimulus. As many studies show, almost any stimulus in an animal's environment can become a conditioned stimulus and produce a conditioned behavioural response. Pavlov's dogs learned to salivate in response to a sound, but they could have just as well been stimulated by a visual signal, like a lamp turning on. Many animals can also easily be conditioned to scent stimuli.

Classical conditioning is most successful when the conditioned stimulus immediately precedes the unconditioned one or the two occur simultaneously. This is no great intuitive leap – if the bell had been rung hours before or after the food was presented in Pavlov's experiments, the dogs certainly would not have formed any association between the two. It is also intuitively understandable that a conditioned reaction gradually becomes weaker and can ultimately no longer be triggered by the conditioned stimulus if the unconditioned stimulus – the reward – does not materialise in the long term. Pavlov's dogs would not have salivated for months in response to the sound of the bell if they weren't given a piece of meat alongside it once in a while.

Conditioned reactions can not only be evoked with a reward or, as they say, positive reinforcement, but also with punishment, or negative reinforcement. If a dog's foot is lightly struck, the dog will raise its paw. If a sound is played while the dog is struck and the whole procedure is repeated several times, after a while the dog will raise its paw in response to the sound alone.

Classical conditioning can also lead to exciting discoveries in animal sense perception. When a Pavlovian dog is conditioned to a tone of 1000 Hz, it not only secretes saliva in response to this frequency, but in response to a 1020 Hz tone as well. This phenomenon is known as generalisation. The more similar the tone is to the conditioned one, the stronger the salivation; the more the tone differs, the less saliva the dog secretes. Logically, then, it can also be concluded that the dog is able to differentiate between frequencies of 1000 and 1020 Hz, or it would have salivated the same amount in response to both tones. If the dog is then only given food in conjunction with the 1000 Hz but not the 1020 Hz tone, after a while it will only salivate in response to the first. This process can be extended, and the tones adjusted to tinier and tinier intervals until a researcher can determine the smallest difference that a dog can distinguish. With the help of such 'conditioned discrimination', as it is technically termed, it is possible to discern what sensory capabilities animals possess and where the limits of their perception lie.

This is the hunch that Karl von Frisch had about a hundred years ago when he set out to test a prevailing theory at the time: that bees could not see colour. In a seminal experiment, von Frisch set out a sugar solution in a glass bowl on a piece of yellow cardboard and quickly found that the bees quickly began to prefer dishes that were on a yellow background compared to those on a blue, green, or purple one. If the bees were conditioned on a blue background, they preferred blue, and so on. Classical conditioning had helped to overturn common knowledge, showing that bees were by no means colour-blind.

Classical conditioning can also show that goldfish can hear. Goldfish will not generally react to the sound of a whistle at the edge of a pond, but if fish food is scattered on the surface of the water, they will swim to the surface and nibble at it. If a whistle is blown before every feeding in the next few days, then the fish will eventually swim to the surface at the sound of the whistle, even if no food comes along with it. Thus, the logic goes, goldfish must be able to hear, or it would not be possible to condition them to a whistle.

Conditioned responses like the ones Pavlov described are common across the animal kingdom, from invertebrates to chimpanzees. Mimicry – the deceptive imitation of signals – is also based in part on this learning process. Birds, for example, are not born knowing that monarch butterflies are partly inedible, but they will quickly find out that eating one leads to vomiting and nausea. When a bird has had this unpleasant experience once or twice, it will avoid all butterflies that look like monarchs in the future. Interestingly, there exists another species of butterfly that look strikingly similar to monarchs but are not toxic, but birds will avoid eating them anyway. By virtue of looking like a 'dangerous' type of butterfly, this non-toxic species benefits from the classical conditioning process that taught birds not to eat monarchs.

ASSOCIATIVE LEARNING: OPERANT CONDITIONING

After classical conditioning, the second most important form of associa tive learning is what is known as operant conditioning. While an animal learns through classical conditioning to link a new stimulus with an existing response, operant conditioning teaches the animal that an initially random behaviour leads to a reward. This form of learning is primarily linked with the name of the American psychologist Burrhus Frederic Skinner, who carried out several studies with rats and pigeons using 'Skinner boxes': devices that contain a lever or a disc that, if pushed, will dispense food on the other side of the enclosure. When a rat is placed inside a Skinner box for the first time, it runs around, explores the enclosure, and interacts with the environment in different ways. At some point it will accidentally push the lever and at a later point it will happen upon a food pellet. After some time, the rat will learn to associate pushing the lever with receiving a pellet and will start to engage in this behaviour for the specific purpose of receiving a reward.

Operant conditioning is therefore also known as learning through trial and error. In principle, this means that behaviours associated with a reward will be repeated whereas behaviours that yield no positive outcomes will be slowly phased out. For an animal to form an association between the behaviour and the reward, the reward must follow the behaviour as soon as possible. If too much time elapses in between, the animal will not learn. Indeed, Skinner noticed a rapid decline in learning success when more than eight seconds passed between when the rat pressed the lever and when it received the food reward.

There are, however, some noteworthy exceptions: if wild rats find food that is foreign to them, for example, they will first eat a very small amount and then wait to see if they feel sick. Even after several hours they can still establish a relationship between eating a certain food and the onset of nausea. But if they do not feel sick, they will start to consume more and more of the new food each following night until they finally eat a normal portion. If they start to feel nauseous at any point during this gradual process, they will still learn to avoid the food altogether.

As this example shows, an animal can in some cases learn to associate a behaviour and its consequence even after a relatively long period of time. What this case study also makes clear is that, through operant conditioning, animals can not only learn how to reap rewards but also how to avoid unpleasant or dangerous situations. As with classical conditioning, the learning success of operant conditioning must always be confirmed over a period of time. Take the rats in the Skinner boxes: once conditioned, they will still press the lever if only every second, tenth, and even hundredth press is rewarded with a food pellet. But if pressing the lever is no longer rewarded at all, the rat will eventually stop doing it completely.

Learning through operant conditioning is extremely important for animals. It plays a decisive role in the search for food, the process of socialisation, the perfection of certain behaviours, and the development of new habitats. Ultimately, this form of learning helps determine which behaviours should be prioritised. Many forms of animal training are also based on operant conditioning.

CAN ANIMALS THINK?

For decades, research on animal learning focused almost entirely on the conditioning process. The neural and molecular bases of conditioning have accordingly been well documented, but the impression has also emerged that conditioning and habituation are the only forms of learning that animals are capable of. Researchers often pushed aside questions of higher cognition in animals, assuming it to be beyond their ability without ever carrying out a study to prove so. That changed completely in 1984 when American zoologist Donald Griffin suggested, in his book *Animal Thinking*, that some animals had consciousness and might have the ability to think. In the years that followed, more and more scientists began to respond to his call to research these mental processes. This line of research experienced such a boom that it led to a new discipline in behavioural biology: animal cognition, or cognitive biology.

Nonetheless, pioneering studies on this topic had been carried out long before Griffin's book. Shortly before the First World War, Wolfgang Köhler became head of the Prussian Academy of Sciences' Anthropoid Research Department in Tenerife. Between 1914 and 1917 he studied 'intelligent behaviour' in great apes under the assumption that they, like humans, can exhibit reasoning and problem-solving skills. An early adopter of film technology, Köhler was one of the first to document the ways chimpanzees use tools. In his most famous study, he placed a banana outside one of his enclosures and filmed the chimpanzee's behaviour. The animals first noticed the fruit and tried to grab it through the bars, but it was too far away. Also in the enclosure were several pipes that the chimpanzees played with from time to time. As the animals grasped for the banana, suddenly Sultan, the smartest of them, took two of the play tubes, inserted the thinner tube into the thicker one, walked towards the bars, and extended the pipe out of the cage to pull the fruit towards him. It appeared that Sultan had recognised a problem and was able to reason through it.

And indeed he had; as a second experiment proved, in which Köhler hung a banana so high in the cage that the chimpanzees could not reach it. At first the animals tried to grasp at the banana as they jumped, but they could not get a hold on it. Several boxes were lying around in the enclosure, and again it was Sultan who took one of the boxes, placed it directly under the banana and began to climb. When he still couldn't reach the fruit, he dragged over a second and a third crate, stacked the boxes on top of each other, climbed onto the rickety structure and was able to jump to the banana. Once again, Sultan achieved his goal through intelligent problem-solving behaviour. From a modern perspective, Wolfgang Köhler's studies represent the beginning of animal cognition research: for the first time, it was shown that animals could learn not only through trial and error, but also through insight.

Still, these findings were ignored for quite a while. It was not until the 1960s when Bernhard Rensch and his team, among others, confirmed Köhler's conclusions at the University of Münster. In a series of elaborate studies, they showed that a chimpanzee called Julia was able to carry out a set of targeted actions after prior planning and consideration. In the first experiment, Julia had learned to throw an iron ring into the slot of a machine, which would then dispense a piece of banana, a grape, or a cookie as a reward. She was then put in front of a small maze over which a pane of glass was laid. Inside was the ring that fit into the reward machine. Julia quickly understood that the ring could be manoeuvred through the corridors of the maze to a side exit with the help of a magnet that was placed alongside the set-up; once she retrieved the ring, she ran to the machine to claim her treats. In the next step, Julia was placed in front of a maze, which consisted of two symmetrically arranged, multi-level compartments, one of which led to the side exit, the other to a dead end. Julia looked at the device for a while, then took the magnet in hand and pulled the ring through the right path to the exit.

Little by little, the scientists increased the complexity of the maze until it contained many turnoffs, angled dead ends, and multiple exits; only one of which the ring could reach. Each round brought a new labyrinth, so Julia had to plan her strategy again and again. The results were astounding: each time Julia studied the maze for about a minute before she began. Then she grabbed the magnet and tried to briskly pull the ring to the exit. She found the right track 86 out of 100 times.

When Rensch and his team carried out the same experiment with six students, they performed only marginally better on average than Julia. The chimpanzee was even *superior* to certain students in some of the tasks.

Many other studies on animal cognition have been performed in the decades since. Köhler and Rensch's findings in chimpanzees were confirmed again and again while other species were also found to exhibit intelligent behaviour. In a simple but ingenious study, five orangutans were given a plexiglass tube about 25 cm long and 5 cm wide that was about a quarter filled with water. A peanut was floating on the surface: one of the apes' favourite treats. Each orangutan's first attempt to fish the nut out with their fingers failed. Yet all five found a solution to the problem: they went to a water fountain outside, took a sip, spat it into the test tube and repeated the process until the water level had risen to a point where they could easily remove the peanut.

Today, no behavioural scientist doubts that animals with highly developed brains such as monkeys, elephants, whales, and apex predators are able to learn through reason. They can grasp a situation, consider a course of action, and then execute it. In other words: these animals can think!

TOOL USE, LEARNING FROM OTHERS, AND CULTURE

Köhler's research not only demonstrated that chimpanzees can solve problems, it also showed that animals can use tools. Sultan's use of the modified pipe to reach the banana outside his cage is one of the earliest documented examples of this ability. Almost half a century later, Jane Goodall described how animals are able to use tools in their natural habitats: wild chimpanzees in the Gombe Stream National Park in Tanzania used stalks, stems, and small twigs to fish ants and termites out of their nests, modifying the plants for better use by plucking off leaves with their mouths or fingers. In other cases, the chimpanzees would create sponges by plucking leaves from plants, chewing them briefly, and inserting the porous mass into small holes in trees that were filled with water. Taking the leaf sponges out again and sucking them dry, they were able to drink. Goodall also observed the animals using leaves to clean themselves and rocks to throw at humans or baboons. More recent studies have also found the chimpanzees to use sticks and stones as hammers and anvils to crack palm nuts. Evidence suggests that these animals have been using this technique in West Africa for millennia and have passed it on from generation to generation. Meanwhile, other species have been found to use tools in their habitats as well: not only great apes, but also capuchin and long-tailed monkeys, sea otters, and dolphins.

If one animal 'invents' a new technique – for example, using a hammer and anvil to crack nuts – then this innovation can spread throughout a population. (Provided, of course, that the species has the ability to learn from role models and imitate their behaviour.) This phenomenon was first observed in a group of Japanese macaques. In 1953, a one-and-a-half-year-old female named Imo invented an unusual way of preparing food on the island of Koshima: she dipped a sand-covered sweet potato in water and wiped the sand off with her hands. A month later, one of Imo's playmates began washing potatoes; after four weeks, Imo's mother began to as well. Four years later the population boasted 15 potato washers, and after 10 years potato washing was a typical behaviour of the entire group. The mother macaques in particular passed this tradition on to their children.

There are now numerous accounts of the cultural transmission of behaviours in the animal kingdom. Such transmission allows different populations of the same species to differ markedly in certain behavioural characteristics, even if their habitats are only a river apart.

These differences have been particularly well-researched in orang-utans. In six populations living in different areas of Borneo and Sumatra, scientists discovered 19 behavioural traits that were most likely passed on through cultural tradition. In one population, almost all the orangutans use tools to catch insects; in the five other populations this never occurs. In some groups the animals build a protective canopy against direct sunlight; in others they do not. Some use leaves as gloves when handling thorny fruits or branches; a behaviour that is unknown to the animals in the other groups. In only one of the populations do the orangutans use leaves as napkins to wipe the latex off their chins. These sorts of behavioural differences do not necessarily have a genetic basis but are likely passed on from generation to generation through social learning.

DO ANIMALS HAVE SELF-AWARENESS?

All the behaviours we have so far discussed would not be possible without highly developed cognitive skills. It is not surprising, then, that over the past few years in behavioural research, questions of animal consciousness have come to the fore. Could it be that a chimpanzee, an elephant, or a dolphin knows what they are? Could they know that other beings also possess unique perspectives? And could it be that these animals are basing their behaviour on this knowledge?

For a long time, it was believed that such questions could not be investigated through traditional biological methods, since observing an animal's behaviour alone does not reveal whether it is based on a higher cognitive performance. After all, good behavioural research requires that a scientist always accounts for the simplest possible explanation.

The renowned Swiss monkey researcher Hans Kummer often illustrated this point in his lectures with the following example: a subordinate baboon is being hunted by a dominant conspecific. The stakes are high for the subordinate: if he is caught, he will be attacked and likely wounded. During the chase, the baboons run by a bush. Suddenly the subordinate stops and stares at it, causing the dominant to pause and look as well. The subordinate then uses this moment to escape. Observing this scene, a researcher is tempted to interpret the subordinate's behaviour as deliberate deception: he pretends to have spotted a danger in the bush, causing the dominant to briefly stop the chase and lose his edge. If that were true, the animals would be demonstrating an extremely high level of cognitive performance that has so far been scientifically proven in only a few species. But there could also be a much simpler explanation: the baboons actually saw something in the bush that the researcher did not.

In fact, there has so far never been an experiment that could prove one animal to have self-awareness and another not to. In the last few decades, however, scientists have developed new methods to approach the question of animal consciousness indirectly. If an animal has self-awareness, it should be able to recognise itself in a mirror, for example: it should see its own face staring back, not that of a stranger.

As early as 1970, the American psychologist Gordon Gallup carried out an experiment with chimpanzees to test this hypothesis. First, he placed a mirror in front of the animals' enclosure for 10 days and watched how the chimpanzees reacted to it. At the beginning they treated their reflections like foreign chimpanzees, threatening the images and screeching at them. But these reactions quickly subsided as they began to use the mirror to get to know themselves better, scratching parts of their bodies they needed the mirror to see, removing pieces of food from between their teeth or making faces.

The chimpanzees' behaviour suggested that they could recognise themselves. Gallup provided the final proof of this with a test he gave on the tenth day: he coloured parts of each chimpanzee's ears and eyebrows red in such a way that the animals could not see the colouring themselves. As long as there was no mirror in the enclosure, the chimpanzees almost never touched the coloured areas; if they did, it was likely by chance. But as soon as Gallup gave them a mirror they immediately began grabbing at the red areas of their bodies. There was no doubt: all four chimpanzees recognised themselves.

Gallup performed the mirror test in exactly the same way with rhesus monkeys and long-tailed and stump-tailed macaques. Remarkably, not a single one of these animals knew who they were seeing in a mirror. This result has been confirmed several times in recent years. In fact, there seems to be a gap between great apes (such as chimpanzees, orangutans, bonobos, and gorillas) and other apes: the former can recognise themselves, the latter apparently cannot. But great apes are not the only animals that can do this: elephants, dolphins, and, surprisingly, magpies all proved capable of self-recognition. Still, most animal species are not.

Even humans cannot recognise themselves in their first years of life: they develop the ability to around one-and-a-half to two years old.

If animals have self-awareness, they should not only be able to recognise themselves in a mirror but also be able to take on another's perspective. In fact, there is growing evidence that some animals are capable of this. In one study, two female chimpanzees who knew each other were set up to compete for food. A dominant and a subordinate, the chimps were placed in two opposite pens, which were separated from each other by a third empty enclosure. In the empty enclosure were two opaque barriers, behind which a researcher placed a plate of food. While the food was always in view of the subordinate, it was sometimes visible to the dominant, sometimes not. In some cases, the dominant could see the food at first, but, while her vision was briefly blocked, the researcher moved the food out of view. In every situation the subordinate had a clear view of the entire process: where the food was, how it was being moved, and what her dominant conspecific was able to see.

The scientists argued that the subordinate, knowing full well which barrier concealed the food, would want to approach it whether or not the dominant's view had been blocked. If the subordinate was aware of what the dominant had seen, however, she would hold back from approaching the right barrier if she knew that the dominant knew where the food was hidden.

When the doors were opened to the middle enclosure, the subordinate chimp indeed behaved as the scientists predicted. As a result, when the dominant had been misinformed or not informed at all, the subordinate was able to get hold of more fruit. Subordinate chimpanzees are apparently aware of what their dominant conspecifics know and adjust their behaviour accordingly.

Many behavioural researchers today assume that great apes actually have a concept of their conspecifics' perception, goals, and knowledge: they can, as they say, put themselves in another's shoes. Recent research with chimpanzees, orangutans, and bonobos shows that these apes even possess another more astonishing skill, which has so far been the claim of humans alone: the ability to know that another holds an incorrect belief and will act accordingly.

What exactly does this mean? Imagine a group of children watching a puppet show. The children watch a robber steal a piece of candy, hide it under a red bucket, and then disappear. Next comes a clown, who looks under the red bucket, takes the candy, hides it under a blue bucket, and leaves. Now the robber comes back to get his candy. If you ask the children between the ages of six and nine where the robber will look, almost all of them will answer 'under the red bucket'. They know the candy is not there, but they also know that the robber holds an incorrect assumption and will act in line with it. Children between the ages of three and four, however, will tell you that the robber will look under the blue bucket: at that age they cannot take on the perspective of someone who believes something that the children know to be wrong.

How can such a study be transferred to animals? An interdisciplinary team from the UK, Japan, Germany, and the USA took advantage of the fact that great apes like to watch television and showed several groups of chimpanzees, bonobos, and orangutans a film in which a person disguised as King Kong steals a stone from a plainly dressed person and hides it in one of two boxes. The undisguised person watches King Kong hide the stone, but King Kong soon threatens the other person and causes him to flee. King Kong then takes the stone, hides it in the other box, waits a moment, and then takes the stone out again and leaves the room with it. The person then comes back to retrieve the stone. While the monkeys were watching this film, an eye-tracker was recording exactly where they were looking. The results were clear: although the animals had seen that the stone was no longer in either of the two boxes, most of them had looked at the box in which the person would mistakenly expect the stone to be. Chimpanzees, bonobos, and orangutans can apparently anticipate that humans will behave in line with incorrect presumptions.

THE SURPRISING COGNITIVE ABILITIES OF BIRDS

Most people assume the smartest animals to be great apes, dolphins, elephants, or apex predators. Until a few years ago, biologists more or less agreed. According to the old dogma, the cognitive ability of each animal species is roughly correlated to the size of their brains and the degree of folding in their cerebral cortex, and these famously intelligent animals are characterised precisely by their large brain-to-body ratio and high degree of cortex folding.

General opinion believes birds to have more 'primitive' brains than mammals, which makes the recent discoveries in avian cognition all the more surprising. Research on corvids and parrots in particular suggests that these animals possess cognitive abilities in line with great apes. Indeed, an African grey parrot called Alex became famous many years ago when his owner, the behavioural scientist Irene Pepperberg, claimed he could understand about 500 words.

When it comes to tool use, corvids are in no way inferior to chimpanzees. New Caledonian crows are some of the best toolmakers in the animal kingdom. In their natural habitat in the South Pacific, these birds form hook-shaped tools from small branches in a three-stage process and cleverly use them as probes to fish insect larvae out of tree holes. In the lab, they have been observed shaping pieces of straight wire into hooks to access a food reward. Rooks have also demonstrated similar levels of intelligence: to catch mealworms floating on the surface of a glass partially filled with water, they gather stones from their lab enclosure and throw them into the glass until the water level has risen to a point where they can reach the food. Magpies, which belong to the corvid group, are able to recognise themselves in a mirror. Corvids are also to some extent aware of the knowledge of their conspecifics: ravens and western bush jays can tell which of their fellow species have observed them caching food and which have not. This is important because members of these species like to plunder other birds' stashes.

Similar to some monkeys and dogs, corvids also have a 'sense of justice' that would be inconceivable without highly developed cognitive skills. Frans de Waal and Sarah Brosnan were able to demonstrate this phenomenon in a classic experiment using capuchin monkeys. First, the animals were taught that they could exchange a token with a researcher for a piece of cucumber; an exchange they accepted with enthusiasm. But if they saw that another monkey received a much more coveted reward – namely a bunch of grapes – in exchange for the same token, then they became indignant and no longer took part in the trade. The animals were even more incensed when they saw that another monkey was given a bunch of grapes without having to exchange a token at all. A few years later, an Austrian research team led by Thomas Bugnyar used the same approach to study ravens and carrion crows and saw the same result: the corvids were also piqued when they saw their counterparts being unjustly favoured.

For biologists the question then arose: how could corvids and parrots have comparable cognitive abilities to monkeys when they do not have a cerebral cortex? The answer can be found in the parallel evolution of the two groups. As far as we know, the evolutionary lines of birds and mammals split from a common ancestor about 300 million years ago. The groups have since evolved in parallel, both developing a large cerebrum, which makes up a majority of the brain. While the cerebrum is organised completely differently in mammals and birds, both forms can produce a high cognitive performance. Bird brains are thus no longer seen as 'primitive' but as simply different. Indeed, the discovery of these animals' astounding cognitive abilities has led to a re-evaluation of the avian brain in the life sciences.

The comparative cognitive performances of different animals have been given great attention by both scientists and the general public. But the enthusiasm for animal intelligence has often neglected a fundamental truth of behavioural biology: while high cognitive ability is the result of natural selection and helps animals adapt to their living conditions, animals with high-level abilities are no better adapted than animals with low-level ones. An animal's adaptive success is shown primarily in its survivability and rate of reproduction, not in its cognitive performance. An earthworm is less intelligent than a raven or a chimpanzee, but it is by no means less adapted to its environment.

CONCLUSION

Learning is common across the entire animal kingdom. Whether simple habituation, like getting used to an irrelevant stimulus, or more complex forms of associative learning, like classical or operant conditioning, learning is a part of the experience of every living thing. Some animals have higher cognitive abilities than others: they can comprehend a problem and plan a solution ahead, they can recognise themselves in a mirror, and they can deduce what their counterparts know, even acknowledging when their counterparts are wrong. Such high capabilities have so far only been shown in a few species; they are by no means universal. Still, the current data show that animals are, quite impressively, capable of cognitive performances that were only ascribed to humans not so long ago. Whether these results indicate that animals possess a sense of self-awareness on par with humans, however, is still a controversial question.

Perhaps the biggest surprise of the last 20 years has been the discovery of remarkable cognition in some bird species. Parrots and corvids in particular seem to be on the same level as great apes. This revelation underscores the fact that evolution towards higher cognition did not only happen in the human lineage; rather, 'intelligent behaviour' has developed independently in the most disparate animal groups.

CHAPTER 6

Animal Personalities

*The Development of Behaviour and the Discovery
of Individuality*

CHILDHOOD AND THE SOCIAL ENVIRONMENT

WHAT HAPPENS TO A MONKEY WHO GROWS UP WITHOUT a mother? Had a biologist been asked decades ago, he or she probably would have answered that if the animal were kept clean, well-nourished, and disease-free it would develop relatively normally. Indeed, until the second half of the twentieth century it was widely believed that the role of the mammalian mother was primarily to provide her children with food (milk especially), warmth, and protection. That she might also play an essential role in her offsprings' behavioural development would have been a rather unusual thought.

The American psychologist Harry Harlow and his team were the first to illuminate the importance of the mother in this respect. In an experiment that by today's standards would be considered unethical, the researchers raised a group of rhesus monkeys from birth with zero contact with their mothers. While all the monkeys grew up to be physically healthy, psychologically they were quite disturbed: some would spend the days squatting on the ground and staring into space; others developed extreme stereotypies, rocking their bodies back and forth for hours. All reacted to new situations with fear rather than curiosity: the sight of a ball, for instance, which would normally entice infant monkeys to play, caused them panic and horror.

When these infants were put in enclosures with conspecifics who had grown up normally, extreme aberrations in their behaviour became apparent: they were hyper-aggressive, uniformly unable to integrate into existing social dynamics. If a female who had grown up without social contact became a mother herself, she did not know how to treat her own offspring. This is of course unsurprising today: it has since been shown that the longer an animal spends isolated from birth, the more pronounced its behavioural disorders and the less successful any attempt to treat them.

Harlow's studies and others have clearly demonstrated how monkeys cannot develop into socially and emotionally competent adults on the basis of instinct alone: proper behavioural development requires intense contact with conspecifics from an early age. The mother not only provides food and protection to her young but she also socialises them, as well as providing security and social support. We have already discussed, for example, how effective mothers can be in buffering their children's hormonal reactions to stressful situations.

Harlow's studies also yielded another important but quickly forgotten finding: monkeys who grew up without a mother but among peers did not develop behavioural disorders. Frequent and high-intensity play with other infant monkeys apparently had a comparable effect to the presence of a mother. Strictly speaking, therefore, it was not the maternal relationship that proved decisive in the animals' development, but the general opportunity for social contact and close relationships with conspecifics.

Research in recent decades has confirmed the importance of socialisation in proper emotional development. Only if infant mammals are socialised successfully will they be able to appropriately interact with conspecifics later in life. This not only applies to primates: dogs, for example, are also heavily influenced by their social experiences between the third and fourteenth weeks of life. Ultimately it applies to all mammals: only when the social environment is intact can infants develop into socially competent adults.

Studies on monkeys, ungulates, and rodents in recent years have revealed another pattern: even quite normal differences in social experience can have a lasting effect on an animal's temperament and behaviour. Temperament is particularly affected by the quality of care an infant receives from its mother: for example, rats that received little maternal care under study did not develop behavioural disorders, but they were significantly more anxious in adulthood than conspecifics who had received a good deal of attention. Mice who received intensive mothering also tended to be more courageous than their counterparts whose mothers had been more reserved.

Still, the mother is not always solely responsible for socialising her child. In many species like the rhesus monkey, contact with playmates of the same age contributes significantly to an infant's development. In other species like the coppery titi, the father is the primary caregiver, while infant elephants are raised by groups of females.

In sum, countless studies have confirmed the importance of the social environment for the behavioural development of young mammals. Individuals can acquire severe disorders in the absence of social partners, and even normal differences in the type of social contact they receive can lead to marked differences in behaviour. The phase from birth through early childhood is therefore considered developmentally crucial to the character of mammalian emotions and behaviour in the long term, although it is not the only stage of life in which behaviour can be shaped by environment.

THE INFLUENCE OF PRENATAL EXPERIENCE ON BEHAVIOUR

When my colleague and long-time collaborator Sylvia Kaiser sat down one morning to look at the footage she had taken of domestic guinea pigs the day before, she could hardly believe her eyes. The four females in the large, richly structured environment she had constructed were somehow behaving like males: they appeared more robust and active than females normally do, performing the courtship ritual known as 'rumba' frequently. Rumba is normally only performed by males – indeed, it is what most obviously sets males apart in large colonies. What was going on?

The second video that Kaiser saw complicated the riddle further. It showed another group of four female guinea pigs living in a separate but identical enclosure behaving in a typical manner for females of their species: they were generally less active, never engaging in rumba or confrontations. How could these clear behavioural differences occur between females of the same age living in identical environments?

Interestingly, there was only one characteristic that distinguished the two groups: the environment in which their mothers had lived when pregnant and lactating. The daughters displaying masculine behaviour were born to mothers who had lived in an unstable environment, while the daughters who were more typically feminine were born to mothers from stable circumstances.

What was the key difference between these two environments? In both conditions, one male lived with five females in a large enclosure. Females from both groups were mated within a very short time and were pregnant for a period of about two months, after which they gave birth to one to four offspring and nursed them for three weeks. In the stable social environment, however, each female had contact exclusively with her mate and the other females of her group. In contrast, females from the unstable environment were occasionally swapped between groups, and thus repeatedly experienced a new social environment with unknown conspecifics.

The mothers of the masculinised daughters had lived in this unstable social environment while they were both pregnant and lactating. Kaiser thus wanted to know whether instability during both phases was responsible for the change in behaviour or if only one of the stages proved critical. Accordingly, she set up enclosures under four different conditions: in the first, mothers lived in a stable environment during both stages, and, as might be expected, the daughters behaved in a typically feminine manner in adulthood. In the second, the mothers lived under unstable conditions during both pregnancy and lactation, and the daughters were accordingly masculinised. Interestingly, daughters whose mothers had experienced instability only during lactation did not exhibit masculine behaviour, while daughters whose mothers experienced instability only during gestation did. The comparison of these four conditions demonstrated that such masculine behaviour stems from instability during gestation, not lactation. Thus, the change is caused by prenatal influence.

Masculinisation, it turned out, was accompanied by significantly increased levels of testosterone in the female guinea pigs' blood. A study conducted at the Netherlands Institute for Brain Research in Amsterdam also showed that the fine structure of certain regions of the brain differed significantly in the masculinised females, appearing closer to the brains of males than their female conspecifics. The social environment in which a mother lives during pregnancy thus influences not only the behaviour but also the hormonal balance and brain development of her daughters.

Consequently, we asked ourselves what effect a guinea-pig mother's social environment during gestation has on her sons. If unstable circumstances tend to masculinise daughters, we reasoned, then it would most likely hyper-masculinise the sons. But the research painted a very different picture. Mothers who experienced social instability during pregnancy produced sons who developed more slowly and were far less likely to express typical masculine behaviour: they played more frequently and at an older age than their conspecifics whose mothers had experienced a socially stable gestation, and repeatedly interrupted courtship behaviour to play once they had reached sexual maturity. Further analysis also revealed a greatly reduced number of testosterone docking sites in their brains. Following these revelations, we termed these guinea pigs 'infantilised'.

Studies on other species including mice, rats, pigs, and monkeys have all produced the same conclusion: mothers who live in socially unstable environments during gestation produce daughters whose behaviour, brain development, hormone balance, and, occasionally, appearance are masculinised; their sons, on the other hand, tend to be developmentally delayed and less masculine in these categories.

Why does the maternal environment during gestation have such a profound influence on behaviour? Although many questions remain, the general way in which prenatal influence functions is now relatively well understood. For one, the environment influences the release of certain hormones in pregnant females. If the mother lives in an unstable social environment in which she frequently has aggressive encounters with unfamiliar animals, her cortisol and adrenaline levels will rise sharply. In addition, sex hormones may also be released. Since the mother's bloodstream is connected to the fetus's through the placenta, the hormones from her body also reach her child's brain. This can have such a lasting effect on brain development that the consequences can be detectable well into the child's adult life.

MASCULINISED DAUGHTERS AND INFANTILISED SONS: DISORDERS OR ADAPTATIONS?

The effects of an unstable maternal environment on a child seem at first glance to be exclusively negative. Accordingly, scientists at medical or psychology conferences tend to present the results from these studies as proof that too much stress during pregnancy can lead to behavioural disorders. Interestingly, however, behavioural biologists tend to discuss these findings in a completely different way. Rather than debating whether the effects of such stress on behaviour can be classified as disorders, we ask: is it possible that a mother is adapting her offspring to her current social situation? In recent years a growing body of research has supported this theory.

Take the common wild guinea pig: like domestic guinea pigs, mothers from this species who experience instability during gestation will produce masculinised daughters and infantilised sons. While these developments seem bizarre at first, under closer examination they begin to look like logical adaptations.

Wild guinea pigs live under severely fluctuating social conditions in their natural habitat in South America. Some years they will live in confined and crowded spaces where they will be frequently involved in confrontations with familiar and unfamiliar conspecifics: clearly an unstable social environment. The very next year, however, extreme weather or predators may decimate the population, and the guinea pigs who are left grow to know and encounter one another in predictable ways. Thus, a mother can be pregnant in two drastically different social environments.

Here's a thought experiment: consider how a pregnant female might optimally shape her daughter's behaviour to a dense and unstable social landscape. What traits would such a daughter have? Assertiveness would no doubt be an asset, as she would often be caught in dominance battles and fierce competition for resources. Although some evidence indicates that masculinised behaviour is associated with an impaired ability to reproduce, at high densities assertiveness is likely to be necessary to bear and successfully rear offspring. But at low densities resources are readily available; being assertive carries no significant advantage. In these cases, behaviour is tailored primarily towards the sort of reproductive potential with which masculine behaviour can very well interfere. Thus, under the ecological conditions of wild guinea pigs' natural habitat, such behavioural changes are necessary adaptations.

Excitingly, a large number of studies over the last two decades on a variety of species have suggested that mothers are indeed able to trigger these adaptations in their offspring. In many cases mothers can influence the behaviour, physiology, and appearance of their young during early developmental stages so effectively that they are indeed optimally adapted to the conditions under which the mothers lived during pregnancy or which they 'predicted' for their offspring.

A particularly spectacular example of this phenomenon can be found in water fleas. One species of water flea occurs in two variations: one with a helmet-like structure on its head and another without. Building such a structure requires a good deal of effort, but in waters rife with predators a helmet provides a certain degree of protection. Interestingly, a mother can 'decide' in the prenatal period how her offspring will be outfitted – if she has had contact with predators, her young will be born with helmets, and if she has lived relatively threat-free they will not.

Mammalian mothers also seem to be able to similarly prepare their offspring for their expected environment. As their brains form in response to the mother's hormonal changes, offspring are able to develop optimally adapted behavioural characteristics.

It is relatively clear why masculinised behaviour would be an advantage for female guinea pigs living in a dense and unstable environment. But what about males? Is infantilised behaviour really an adaptation for them, too?

For all we know, it certainly is, given that the reproductive path adolescent males take depends completely on the population density of their environment. Imagine a population consisting of only two young and sexually mature males and one young female. How should the males behave to successfully pass on their genes? They should attack and try to defeat their opponent of course, as only the dominant male will be able to mate. Like many other mammals, this is indeed how guinea pigs behave. Infantilised behaviour would clearly not promise success in this situation.

At high population densities, however, the circumstances are completely different: a young male must constantly face older, stronger alpha males who ardently defend the females with whom they mate. The young males could theoretically reproduce, but since they are much smaller than the alphas, they have no chance of defeating them. As we saw in chapter two, when the population density is high, adolescent males attain alpha status by forming bonds with individual females and investing all their available energy into gaining body mass. They also cannot reveal themselves as serious competitors and risk being challenged by the current alphas, so infantilised behaviour is the perfect strategy: the young males are able to signal to the alphas that they are not rivals, and they do not waste their energy on fights they would lose anyway. Studies have shown that this approach is quite successful in helping the young males achieve alpha status themselves when they are larger and heavier.

In sum, there is no research that suggests that masculinised daughters and infantilised sons are behaviourally disturbed. In fact, most evidence indicates that these animals are optimally adapted to certain social situations.

ENVIRONMENT, GENES, AND SELF-INTEREST IN THE EARLY DEVELOPMENTAL STAGES

We can now clearly see how the social environment triggers hormonal changes in pregnant females and influences the brain development and behaviour of their offspring. But environmental conditions can also affect the ways in which mothers care for their young after giving birth: mouse mothers, for one, will give a great deal of attention to their young in a safe environment, but hardly any in a dangerous one; bighorn sheep will similarly care intensively for their young only at low population densities. The differences in these mother–child relationships also have marked effects on the brain development of offspring, which again results in adapted behavioural patterns: children from less present mothers tend to behave more cautiously, for example, which certainly makes sense for survival in a dangerous environment.

Indeed, there is a remarkable correspondence between the pre- and postnatal phases: in both stages of life, the maternal environment is reflected in the brain development and adapted behaviour of the off-spring. The major difference is that these changes are caused by hormones during gestation and by the mother's behaviour after birth.

But while the social environment during these early stages is incredibly influential to a child's behaviour, it by no means shapes the child alone. Children are not passively socialised: they play a highly active role in their own development.

As the American evolutionary biologist Robert Trivers pointed out decades ago, children and parents often have different interests. Because natural selection programmes each individual to pass on as many copies of its genes as possible, each child will try to extract a maximum amount of food and attention from its mother. The mother, meanwhile, intends to distribute resources equally among her offspring, since expending too much time on an early litter will prevent her from raising another. This inevitably results in conflict between a mother and her young, as each child wants more from the mother than she is willing to give. The mother and other caregivers thus by no means shape young animals' behaviour alone: the offspring themselves also exert a considerable influence on their conspecifics. We have seen how successful they can be in convincing others to care for them through the baby schema, which can cause even unrelated conspecifics to affectionately care for another's young.

If all mothers in a population can adapt their offspring to certain conditions, it follows, then, why don't all infants develop in exactly the same way? Why do identical environments produce completely different personalities?

The fourth chapter already provided a partial answer to these questions: individual behaviour is always shaped by the interaction between genotype and environment, a dynamic in which plasticity genes like SERT play an essential role. Such genes can occur differently across individuals and, depending on the variant, determine the extent to which an animal's behaviour is influenced by its environment. Thus, the same level of maternal care does not necessarily produce the same behaviour in different offspring: reduced maternal behaviour generally leads to more anxious children, but the degree to which that anxiety is expressed is determined, among other things, by the child's SERT gene.

THE EFFECT OF ADOLESCENT EXPERIENCES ON BEHAVIOUR

Behavioural science had previously long believed that the patterns that emerge during a mammal's early developmental stages last throughout its life: fearful infants grow into fearful adults; brave ones carry their courage with them as they age. As research progressed, however, scientists began to wonder: how decisive are the early phases in determining behaviour? Is it not possible for the environment in an animal's later years to be equally formative? In fact, recent studies have broadened our view of behavioural development, increasingly placing the focus on adolescence.

Adolescence is defined as the gradual transition from childhood to adulthood. It is marked by profound hormonal changes: in females, the ovaries begin to produce sex hormones like oestradiol, viable eggs begin to mature, and ovulation occurs for the first time; in males, the testes activate and viable sperm and hormones like testosterone are first produced. Sex hormones also tend to cause significant changes in appearance: male mandrills develop colourful faces, for example; male deer grow antlers; and roosters develop a bright red comb. The females of many monkey species, meanwhile, start to show conspicuous swelling in their anal and genital areas. The nervous systems of many animals also start to undergo serious changes during this time, as certain circuits are reorganised and sex hormones start to enter the brain via the bloodstream and dock in areas responsible for emotions and behaviour. Amid so many physical transitions, it is not surprising that animal behaviour also drastically changes.

As the child's need for its parents decreases, its interactions with peers become increasingly important. Hormones awaken interest in the opposite sex in both males and females, while males often become less compatible with each other and migrate to separate social groups or establish their own territories. They also begin to display a new willingness to take risks and search for new and exciting experiences: characteristics that can be attributed, among other things, to an increased concentration of testosterone.

Behaviour also responds in step with these hormonal changes, but not in the same way in all mammals: just as in the prenatal and early childhood phases, the environment during adolescence plays a role in shaping future behaviour. Again, our studies on domestic guinea pigs were the first to demonstrate the crucial role of social experience during adolescence in the development of behavioural patterns.

Consider a male guinea pig who has spent his entire life in a large colony. What would happen if he suddenly encountered an unfamiliar male from another colony in an enclosure that was unknown to both of them? The two would likely look at each other, take a few sniffs, and then peacefully determine the dominant. The arrangement would proceed stress-free, as studies measuring cortisol levels have shown. Likewise, as we have seen in chapter two, males raised in large colonies have no issues integrating into new ones.

But such encounters play out quite differently with males raised individually or in pairs, who will generally chatter their teeth, raise their hackles, and charge at one another – sometimes for days – until a dominance relationship is finalised. These encounters cause a sharp rise in cortisol levels for both animals, which only gradually fall back to baseline. Unsurprisingly, these animals also experience violence and stress when they try to integrate into unfamiliar colonies during adulthood.

How do these differences arise? Why is meeting an unfamiliar conspecific relatively relaxed for male guinea pigs who were raised in colonies, but highly stressful for males who were not? We conducted a series of studies to find the answer: the social experiences guinea pigs have during adolescence shape their ability to socially integrate. Young males growing up in colonies learn how to interact as an inferior with larger, more dominant alphas daily, and simultaneously encounter younger conspecifics whom they can potentially dominate. They also figure out how to deal with females, learning how to wait about three hours for a dominant male to appear before beginning to court anyone who catches their interest. This is extremely important for successful integration. Competition for females is one of the main causes of aggressive encounters between males in the animal kingdom, and certain learned behaviours prevent conflicts from escalating.

We can see the value of such experiences by looking at the integration skills of males raised alone or with a single female. Having had virtually no opportunity to interact with others during this crucial developmental phase, these males never learned the rules of subordination or other important social skills, so their confrontations with unfamiliar males are aggressive and highly stressful. Social experiences during adolescence, we can see, are essential for learning how to regulate conflict.

But what about females? Do social experiences during adolescence also influence their adult behaviour? Unfortunately, hardly any studies have provided a reliable answer to this question. But preliminary research on domestic guinea pigs has yielded some surprises: in stark contrast to males, females seem to get along with unfamiliar conspecifics regardless of their previous experiences and can easily integrate into unfamiliar colonies.

The effect of adolescent social experience on male behaviour has been confirmed by studies on monkeys, apex predators, other rodents, and, interestingly, zebra finches: males of this species that live in pairs during adolescence become extremely aggressive towards strangers later in life, while conspecifics from colonies tend to be much more relaxed. These findings on a wide variety of species no doubt lead to the general conclusion that behaviour is shaped not only by the environment during an animal's early stages of life, but also by its social experiences during adolescence.

The extent to which later behaviour actually reflects these experiences, however, is generally determined by genotype. We have already seen how this happens during childhood: animals who are similarly socialised can react completely differently to the same event depending on their genes. The few studies that have explored this phenomenon during adolescence have yielded similar conclusions. One study on mice, for instance, found that if males experience defeat in a confrontation during this period they tend to behave more anxiously in the future. But the extent to which the experience changes their behaviour depends largely on the animal's SERT gene: those with two defective alleles are significantly more fearful after defeat; those with two intact alleles are hardly bothered; and those with one intact are affected at a level somewhere between the other two genotypes.

ADOLESCENCE: A PHASE OF ADAPTATION

Most of us would consider an animal who gets along peacefully with conspecifics to be more likeable than one who is highly aggressive towards strangers. Given the choice between two dogs of such different natures, for example, most of us would choose the peaceful one. In guinea pigs, too, it is reasonable to conclude that males are better socialised in large colonies than in isolation or pairs.

But evolutionary biology has a different perspective: the question is not which type of behaviour is more palatable to humans, but which type is more likely to lead to reproductive success. As we will see, the answer depends entirely on an animal's current social environment.

A more peaceful behavioural profile perfectly suits a male domestic guinea pig attempting to integrate into a large colony, as it allows him to bide his time until he can reach alpha status and reproduce. An aggressive and stressed male would have little chance surviving in such a situation, but a decent chance of being successful in a pair. Why? Because male guinea pigs who grow up with one female are able to regularly reproduce and live contentedly. Suppose a rival suddenly appears: how should the male react? From an evolutionary perspective, the best course of action is to attack the intruder as defeating him is the only way to guarantee the male's continued reproductive success. The second chapter has already laid out how stress hormones enable an animal's body to react quickly and effectively to threatening situations. Thus, high levels of aggression and cortisol are advantageous to a solitary male as they help him win a fight. Highly aggressive behaviour in a pair situation is therefore just as perfect an adaptation as peaceful behaviour in a large colony.

My colleague and collaborator Tobias Zimmermann demonstrated in his dissertation how an aggressive-stressed profile is indeed an advantage in deciding reproductive success through confrontation. In his study, Zimmermann formed groups of guinea pigs consisting of two females and two males. One of the males in each group had been raised in a large colony while the other had been raised with a single female. As might be expected, the males from pairs were more aggressive and stressed than their rivals from colonies, attacking their opponents immediately and dominating them in most cases within a few hours. Females showed no preference for either male initially, but subsequently orientated themselves almost exclusively towards the dominant males after these confrontations. As genetic paternity studies showed, the aggression paid off: the males from pairs sired significantly more offspring than their peacefully socialised competitors.

Thus, when it comes to reproductive success, neither peaceful nor aggressive behaviour is inherently more advantageous: both are optimal adaptations to particular situations. Aggression tends to be more effective in fierce competition for territory or mating partners, while in more complex hierarchical systems, peaceable behaviour is more rewarded. These patterns are not innate: they are adaptive strategies learned during adolescence.

Adolescence is generally considered to be the final stage of life in which behaviour is shaped. Unlike the prenatal phase (during which animals cannot perceive the environment with their own senses) and infancy (when their behaviour is shaped primarily by their parents), adolescence is the first time when an animal seems to ask itself: Did my mother teach me the right way to behave? Am I actually well adapted to my environment?

This phase of questioning is quite useful as the environment cannot always be predicted during the early developmental stages. Living conditions may change as an animal is growing up: a stable social environment may destabilise, and the number of conspecifics with which an animal interacts may increase or decrease dramatically. If an adolescent's current behavioural profile proves to be unsuitable, it can still change fundamentally – perhaps for the last time. More importantly, certain personality traits tend to emerge during this phase, like an individual's tendency to be peaceful or aggressive towards strangers. Adolescence is a sensitive period of life during which large and small behavioural adjustments are still possible, thus adapting an animal to the social environment in the best possible way.

The current state of the research holds that behavioural development is a process extending from gestation through adolescence. Consequently, we must ask: Do other sensitive phases exist later in life during which environmental influences can have formative effects on behaviour? As far as we know, mammals can change drastically in response to the environment well into adulthood; thus victories or defeats in the struggle for territory, ascents or descents in the social hierarchy, and the gain or loss of social bonding partners all bring about strong changes in hormonal balance, neural circuitry, and behaviour. Animals even continue to learn throughout life: this applies to a 12-year-old German shepherd, a 30-year-old dolphin, and a 50-year-old elephant alike, although learning is more difficult in old age than in childhood. The most determinative years for behaviour are up until the end of adolescence, but change is still possible throughout life.

THE DISCOVERY OF INDIVIDUALITY

From all we have heard so far, it is not surprising that adult animals of the same species differ markedly in their behaviour and temperament. Indeed, animals form unique characteristics – we can even call them personalities – in response to genetic predispositions and social experiences at different developmental stages. The study of such 'animal personalities' has become one of the most exciting and cutting-edge focuses of behavioural biology in recent years, producing about 9000 scientific articles since the last turn of the century.

Research from all species studied so far has confirmed that animals from the same populations consistently differ in temperament and behaviour over long periods of time. If animal A is more courageous than animal B today, for example, it generally was four weeks ago as well, and will continue to be so four weeks into the future. At first glance these findings do not seem particularly exciting; in fact, they are what we would expect from our intuitive understanding of chimpanzees, dolphins, elephants, and many other species. Every dog owner knows how Luna is different from Emma, and Henry is different from Baloo. The spectacular thing is that this finding extends to wild populations of songbirds, fish, reptiles, and even insects, and that such differences between animals are permanent, and indeed are suggestive of distinctive personalities.

Let's take a closer look at one example from the research, in which a group of scientists studied populations of great tits in Belgium, Germany, Holland, and the UK over a period of many years. To observe the behaviour of different individuals in response to environmental changes, the researchers took certain birds from their natural habitat and placed them in a large aviary the next day, where they would fly around, hop from branch to branch in the five available trees, and generally orientate themselves to the unfamiliar terrain. Observations of the animals' exploratory behaviour showed great differences between individuals: some birds were quite courageous and explored the new environment with enthusiasm, while others were more hesitant, reserved, and shy; others still behaved somewhere in between. After the team had completed their observations, each bird was released back into its natural habitat.

A few months later, the research team recaptured several of the great tits that they had previously studied and once again examined them in a new environment. Now they observed something remarkable: the courageous birds were once again exploring the aviary bravely while the hesitant ones were still holding back. In other words, the differences in personalities had remained consistent over several months. New research has shown that such personalities can even be observed in invertebrates. Leaf beetles, for example, differ markedly in how quickly and actively they explore foreign terrain: if studied under similar circumstances to the great tits, the same type of consistent temperamental differences are apparent.

Until a few years ago, studies like these would have mainly focused on whether animals in the UK are more eager to investigate a new environment than those in Belgium, for example, or whether those in a forested area in the Netherlands are on average braver than those living by a lake. The research team would have calculated a mean value of 'keenness to explore', which included the variance around the average. The variance would have been evaluated as noise: irrelevant data. But this view has now fundamentally changed: the variance is no longer seen as noise, but as an expression of individuality. The individual is now the centre of scientific interest.

Research into animal personalities has also yielded another insight: certain aspects of behaviour are often tightly intertwined, resulting in what are termed 'behavioural syndromes'.

A study on sticklebacks conducted by a research team at the University of California, Davis, provides a good example here: first, the researchers examined a shoal in an unfamiliar aquarium and took note of how eagerly each fish explored it. Some were highly active, others were quite passive, and others behaved somewhere in between. In a second step, they observed how the same fish behaved towards foreign conspecifics, and again saw a large variety of reactions ranging from highly aggressive to completely calm. Finally, they used a third test to measure how quickly the same sticklebacks began to feed again after a simulated heron attack. Here too, they saw a spectrum of courageous animals who began feeding again relatively quickly to more fearful ones who completely lost their appetites.

Interestingly, the different types of behaviour that the fish exhibited showed certain correlations: if a stickleback eagerly explored the new environment, it would also behave aggressively towards new conspecifics and quickly start feeding again after the simulated attack; if a fish was shy in the new aquarium, it was also peaceful with strangers and hesitant to begin feeding again. Aggression was apparently coupled with courage, and passivity with restraint. In fact, numerous studies on a wide variety of species show that such links between types of behaviour frequently exist.

But research into animal personalities has also challenged our ideas about the mutability of animal behaviour. Behaviour has traditionally been viewed as a trait so flexible that it can be changed nearly at will to any situation. Natural selection, the logic follows, should promote animals who are able to behave optimally in as many circumstances as possible, and the most successful animals should accordingly be those who can behave boldly when necessary, but just as easily retreat when hesitance proves to be an advantage. But, as we have seen, animals do not behave this way; instead, they demonstrate how limits exist to an individual's behavioural adaptability. Like humans, sometimes animals simply are the way they are. They have characters and personalities, which are expressed throughout their lives and which cannot be changed at will.

Why is this? It apparently takes considerable effort to modify a personality once it has solidified: neural circuits must be switched off and formed anew; systems of hormonal control must be readjusted. All of this requires time and energy. Behavioural biology thus talks about the 'costs' of flexibility, which ultimately result in the fact that animals do not adapt their behaviour to every new situation, but rather develop relatively stable behavioural profiles. No profile is optimally adapted to all situations: bold personality types have an advantage in the hunt for food, but a disadvantage in avoiding predators; aggressive types may be successful in confronting rivals, but less so in finding mates. Evolutionary biology holds that different types can coexist in the same population, but only if they have comparable reproductive success: natural selection ultimately favours animals who are more successful in passing on their genes. Thus, if a certain personality type has little-to-no reproductive success, it disappears from the population.

We recently investigated a different but no less exciting aspect of 'individuality' with an interdisciplinary team of behavioural scientists, neurobiologists, psychologists, and computer scientists. As we have discussed, it is widely accepted that the distinct behavioural profile of every animal and human arises from the interplay between genetics and environment. But what happens in a population of identical genotypes living in the same environment? Do stable individual personalities also arise under these conditions?

To answer this question, we placed 40 mice in a richly structured, barn-like enclosure much like the one described in chapter five. All mice were female, four weeks of age, and carrying the same genes. They were also all fitted with tiny chips, which would interact with antennas around the enclosure to signal when other mice approached. We then used a database to determine which mice were where in real time, and we recorded the movements of all the animals day and night over a period of three months.

The analysis of this massive data set at first showed hardly any difference in how the individual mice explored their environment. Gradually, however, different character traits emerged: some mice were highly active and could be found almost anywhere, while others stayed in certain places most of the time, and others still showed patterns somewhere between the two extremes. Over the three months, each animal began to show a stable, characteristic behavioural pattern. This study led to the spectacular realisation that even genetically identical animals from the same environment tend to develop distinct behavioural profiles.

CONCLUSION

The shaping of mammalian behaviour is a process that extends from the prenatal period through infancy and adolescence. Throughout these phases, the social environment plays an essential role: the mother in particular influences the brain development of her offspring during their early life stages and adapts their behaviour to the environment. But genetics is also hugely important in shaping the way individuals react to the behaviour of their mothers and other social partners. Characteristic behavioural patterns thus arise from the interplay between genetic predisposition and environment, and remain dynamic until adolescence, when an animal's behavioural profile can still be fundamentally changed – perhaps for the last time – by social experience. Once an individual has entered adulthood, it should be adapted to its environment in the best possible way.

Research in recent years has begun to focus on unique animal personalities. Although many questions remain, we can already say with confidence that diversity among individuals is a basic feature of behaviour: indeed, we can only reach a comprehensive understanding of animals if we take this insight into account.

CHAPTER 7

Altruistic Squirrels and Egotistical Lions

The Sociobiological Revolution

I N 1975, THE AMERICAN BIOLOGIST EDWARD O. WILSON published his foundational work, *Sociobiology – The New Synthesis*. In it, he coined the term 'sociobiology' as a new sub-discipline of behavioural biology based on the groundwork laid years earlier by scientists like William Hamilton and Robert Trivers.

Wilson saw the goal of sociobiology as deciphering the biological basis of all social behaviour of animals and humans alike. By systematically applying the theory of evolution to the social activities of insects, fish, birds, and mammals, sociobiology casts the ways in which animals form relationships, help one another, kill each other, and take on sex roles in a completely different light.

Behavioural biology was initially hesitant about this new theory, but soon became enthusiastic, accepting it as an explanatory approach to the function and evolution of animal behaviour. By explicitly including humans in his considerations and provocative theses, Wilson also aroused the interest – and soon opposition – of the social sciences. A public debate soon arose around the term 'sociobiology' and the extent to which (if at all) it can provide comprehensive explanations for human behaviour. But these debates will not be the subject of this chapter. Rather, we will focus on what the sociobiological revolution has done to our understanding of animal social behaviour. To do that, we must first turn back to Darwin.

DARWIN'S PROBLEM

Darwin identified two underlying factors of biological evolution in *On the Origins of Species*: First, all animal and plant species must contain differences between individuals that are at least partly hereditary, and second, individuals must differ in their reproductive success. The human species illustrates his first point well: people can be tall or short, blonde or brunette, heavy or light, brown- or blue-eyed. These differences are partially based on genetic inheritance, but, depending on the characteristic, they can also be greatly influenced by environmental influences. Eye colour, for example, is purely an expression of genetics, while weight is significantly affected by environmental factors like nutrition. Only the hereditary share of such traits plays a role in biological evolution.

As for Darwin's second point, behavioural research on a large number of species has indeed proven that individuals within a population differ in their reproductive success. For instance, in one famous study, the British zoologist Tim Clutton-Brock and his team looked at a population of red deer on a small Scottish island over many years. While the females in the group gave birth to an average of four to five viable calves over their lives, some individuals gave birth to up to 13 offspring, while a third had no reproductive success at all. The differences in success between males were even greater: some sired up to 24 offspring, while more than 40 per cent sired none.

As Darwin recognised, the correlation between these two factors – heritable differences in traits and individual reproductive success – leads to long-term changes in the genetic composition of a population. Thus, if an animal with certain genetically determined traits leaves more viable offspring than others, the traits of the reproductively successful individuals will be more present in the population. This dynamic is precisely what characterises the evolutionary process.

But what determines such drastic differences in numbers of surviving offspring? Darwin's answer had to contend with two facts that were well-known at the time: first, that every animal species gives birth to far more offspring than are necessary to maintain the next generation, and second, that in the vast majority of species, the size of a population over generations remains relatively constant.

Darwin resolved this apparent contradiction. Most offspring perish, he recognised, while only a few survive to sexual maturity and even fewer subsequently reproduce. For instance, the adult females in a Pacific species of salmon produce about 6000 eggs, all of which are fertilised by a single male. But if the number of individuals remains roughly the same across generations, then these 6000 fertilised eggs must produce on average only one female and one male (who each reproduce at the age of five). The same is true for herring gulls: a successful pair typically produces three eggs per breeding season – about 30 eggs over the course of their lives – although statistically only two of these eggs develop into adults who go on to reproduce. The greatest losses in offspring occur before sexual maturity; in some populations, 25 per cent of laid eggs do not even hatch. Of those that do, 40 per cent tend to die before fledging – most in the first week of life – and another 40 per cent do not survive the winter.

Darwin discovered that the pattern of which animals survive, reproduce, or perish is by no means a matter of chance. Those who survive are genetically better adapted to their environment: they are better able to recognise predators, make use of food, effectively court mates, and care for their young, and in turn they are more likely to reproduce than their conspecifics. Thus, the genetic advantages that allowed these individuals to reproduce in the first place are preserved in the population, while the genetic material of their less-successful counterparts is not.

This process of natural selection constantly adapts the animals of a population to their environment and orientates their behaviour towards the goal of transmitting their genes to the next generation. In other words, the ultimate goal of all animal behaviour is to maximise individual reproductive success – or, in evolutionary terms, to maximise Darwinian fitness.

But for many decades this theory had a huge problem. it implies that natural selection 'programs' an animal to behave egotistically, acting only in the service of passing on its *own* genes. But if we look at the animal kingdom, we quickly come across examples that seem to contradict this theory. Many bees and wasps and all ants, for example, live in colonies in which individuals are assigned to different castes to fulfil different functions: namely, workers defend and take care of the colony while the queen reproduces. Remarkably, the workers themselves are sterile. And therein lies the problem: how can natural selection favour individuals who do not reproduce? It undoubtedly benefits the queen and her offspring to be fed and defended by the workers, but the advantage for the workers themselves is quite unclear. The warning calls of many birds and mammals are another curiosity: when an animal detects a predator, it sounds an alarm call, prompting all its conspecifics in the vicinity to move to safety. But in doing so, the animal who sounded the warning draws attention to itself and increases its own risk of being killed. This behaviour, which at first glance appears altruistic, does not seem compatible with Darwin's theory of evolution. If maximised personal fitness is every organism's goal, why don't these birds or mammals simply run away from the predators they sense? Why do they try to increase the survival of their conspecifics by putting themselves at risk?

To add a further example, many mammals even engage in communal suckling. Mice and lions, for instance, not only provide milk for their own babies but also for the offspring of other females. Why? If animals are only preoccupied with helping their own young to survive, why do they share their resources with the children of others?

Remarkably, Darwin himself clearly recognised that his theory could not adequately account for such seemingly altruistic behaviour. The question of how natural selection could have produced the sterile caste of colony-forming insects particularly troubled him. Although he never came up with a satisfactory solution, he came very close by assuming that the phenomenon has something to do with family relationships.

OF ERRORS AND LEGENDS

Several decades after Darwin, Konrad Lorenz and most of his contemporaries no longer had trouble explaining such altruistic behaviour. Unlike Darwin, these scientists assumed that animals behaved for the good of the species rather than their own personal reproductive success. Simply put, it did not matter precisely who lives and who dies, who escapes and who gets eaten, who reproduces and who looks after the pack – it was only important that the species continue. This idea was still widespread in science until the 1990s, and still today it is ubiquitous in public opinion and popular science reporting.

The current scientific consensus, however, holds this view to be false. Why? The American cartoonist Gary Larson gave the argument most succinctly: one of his cartoons depicts lemmings trying to throw themselves off a cliff into the sea, referring to the myth that these animals commit mass suicide for the good of the species to provide the few that live on with sufficient resources to survive. This myth is based on reports that these small rodents will migrate en masse when the population becomes too large and resources too scarce.

So far, so good, but one of the lemmings in Larson's cartoon is clearly different: he is wearing a life preserver. And this is precisely the point: acting solely for the good of the community only functions until one animal begins to behave selfishly. All the unselfish lemmings will run off the cliff, and their genes will go with them. The egoist, on the other hand, will survive and reproduce, meaning, in a sense, that his sons and daughters will also wear life preservers. Thus, the genes that code for selfless behaviour must disappear sooner or later. In technical terms, this means that selflessness is less evolutionarily stable than selfishness.

So, it is unsurprising that lemming mass suicide has proven to be a myth. It is true that the population size of these small rodents from the northern Arctic regions is subject to regular cycles of three to four years, during which the number of individuals multiplies more than a hundredfold and then quickly collapses, leaving only a few animals in the area. But there is no evidence that these cycles are driven by mass suicide. Studies have shown that populations of collared lemmings, a species endemic to Greenland, are primarily decimated by predators like stoats, whose numbers increase distinctly in response to rising numbers of lemmings.

For other species of lemmings living in Alaska or Scandinavia, it is also quite conceivable that food shortages lead to increased migrations, sending local population sizes into sharp decline. Such migrations are not without risk: death is possible at every turn. When the lemmings must cross a body of water, for example, it is almost certain that at least one will drown. Anecdotal observations of such events could very well be the basis of the mass suicide legend. The myth was certainly popularised by the 1958 Disney movie *White Wilderness*, which purportedly documented the migration of lemmings. But the famous leap over the cliff did not naturally occur: it was staged by the studio.

THE MEANING OF KINSHIP: THE ALARM CALL OF BELDING'S GROUND SQUIRRELS

Although lemmings jumping off cliffs en masse has proven to be a legend, researchers have discovered many other examples of such seemingly selfless behaviour. All scientifically verified, these include communal suckling, warning calls, and sterile castes. But how can such behaviour arise if it is not evolutionarily stable? In the mid-1960s, British biologist William Hamilton provided an essential answer with his theoretical work highlighting the importance of kinship for the behaviour towards conspecifics.

This phenomenon can be nicely illustrated by an impressive study conducted by the American scientist Paul Sherman on Belding's ground squirrels. This species of ground squirrel lives in large colonies in the mountainous regions of the western United States, where they must constantly be on guard for birds of prey, badgers, coyotes, martens, and weasels. If an enemy on the ground approaches, the squirrel who perceives it sounds an alarm telling its conspecifics to move to safety. Sounding the call carries a known risk: in Sherman's study, nearly 10 per cent of the animals who gave the warning were eaten by predators, while 50 per cent of overall victims had warned of a predator shortly before. When Sherman further analysed the data, he made an even more remarkable discovery: not all animals equally took on the risk of warning the others. Adult females made alarm calls disproportionately often, while the adult males tended to hold back.

Sherman was well acquainted with the kinship ties of the animals, and thus knew that females are usually surrounded by other female relatives throughout their lives: grandmothers, mothers, daughters, granddaughters, aunts, sisters, and cousins typically bound together in clans. In contrast, adult males in a certain area do not tend to be related to each other or to females, and thus are constantly surrounded by non-relatives. These divergent patterns stem from differences in migration behaviour: when young males reach sexual maturity, they migrate to more distant regions, while daughters tend to establish themselves close to their natal burrows.

Could the females be more likely to sound warning calls to protect their relatives? Do the males tend not to warn their nearby conspecifics precisely because they are not relatives? To test these hypotheses, Sherman compared the behaviour of females who were surrounded by relatives to that of females who had no living relatives left. Indeed, females who were close to their own offspring gave significantly more warning calls than females who were not near their children. Even females who only lived close to a mother or a sister gave significantly more warnings than females with no relatives nearby at all. Thus, females adjusted the frequency with which they warned of a predator to the number of relatives in their vicinity: the more close kin lived nearby, the more frequently they sounded alarm calls. The risk of the behaviour was taken on to benefit a certain lineage.

Countless studies on a huge diversity of species have confirmed that kinship has a decisive influence on how animals behave towards conspecifics. As a rule, animals tend not to demonstrate altruistic behaviour towards anyone but close relatives.

WILLIAM HAMILTON AND KIN SELECTION

But why is kinship so important to the evolution of social behaviour? As mentioned, William Hamilton's theoretical investigations had already found the answer more than 10 years before Sherman's studies. Hamilton began with Darwin's insight that an animal's tendency to feed, defend, or put itself in harm's way for its young is basically unsurprising. Such seemingly selfless behaviour is indeed selfish from an evolutionary point of view: offspring carry their parents' genes, and thus parents continue to ensure their own reproductive success by protecting and nurturing their young, even to their own detriment. In other words, through altruism towards their kin, parents increase the likelihood that their own genes will be preserved in the next generation.

Hamilton considered relatives on par with children as far as altruistic behaviour is concerned, as they likely also possess close copies of an organism's genes. In animals like humans who carry a double set of chromosomes, each gene occurs in two alleles. Every individual receives half its alleles from its father and the other half from its mother, and accordingly passes half of its own alleles to each of its children. There is a 50 per cent probability that a certain allele – eye colour, for example – is

ALTRUISTIC SQUIRRELS AND EGOTISTICAL LIONS

passed from parent to child, and thus a 25 per cent probability that that same allele is passed from grandparent to grandchild. The same principle can be used to calculate these probabilities for other relatives: nieces and nephews, for example, have a 25 per cent chance of carrying the same alleles as their aunts and uncles, while cousins have a 12.5 per cent chance of sharing an allele from a common ancestor. For siblings the probability is 50 per cent, and for identical twins, 100 per cent. In general, the higher the degree of relationship between two individuals, the more identical alleles they possess.

These probabilities amount to the fact that an organism has the same chance of transmitting its own genes whether it produces offspring itself or helps a sibling care for two of its own: in terms of how many copies of the animal's own genes make it to the next generation, one child is equivalent to two nieces or nephews. Accordingly, an individual with one of its own children and three nieces has more of its genes in the next generation than an individual with two of its own children and no nieces or nephews.

The evolutionary definition of 'fitness' – the contribution an individual makes to the gene pool of the next generation – can admittedly lead to misunderstandings outside of biology. Hamilton recognised that fitness is made up of two components: the proportion of genes that an animal passes on to its own offspring, which is known as 'direct' or 'individual' fitness, and the proportion of its genes that are present in relatives and passed onto their offspring, which Hamilton termed 'indirect' fitness. Accordingly, evolutionary success ultimately does not only depend on an individual's direct fitness, as Darwin thought, but its 'inclusive fitness': the sum of the direct and indirect components.

It thus follows that individual A with one child and three nieces has a higher indirect fitness than individual B with two children and no nieces or nephews. The three nieces increase the individual A's inclusive fitness as well, while the inclusive fitness of the individual B is equal only to its direct fitness since all copies of its genes in the next generation are present in its own offspring. Which variant would prevail in a population: A or B? The answer is A. Natural selection ultimately favours the variant that passes on its genes to the next generation with the highest efficiency, meaning the variant with the highest overall fitness will prevail.

So, forgoing one's own offspring to help relatives care for theirs can indeed prevail over the course of evolution, provided that this altruistic behaviour ultimately leads to a higher overall fitness for the individual than rearing its own offspring would. Or, in terms of Hamilton's famous formula: altruistic behaviour can evolve when the cost to the altruist is lower than the benefit to the individual receiving the support multiplied by the degree of relation between the altruist and the recipient.

Altruistic behaviour should therefore be directed not only at one's own children but also at the reproductive success of relatives. This of course means that such behaviour does not turn out to be so altruistic after all: it is in no sense selfless, but an effective means of genetic preservation.

It bears pointing out that animals do not consciously consider their own relative closeness to certain conspecifics, nor do they make calculations about which behaviour will benefit their direct or indirect fitness. Rather, they are 'programmed' by natural selection over generations to behave in ways that maximise their total fitness. But how do animals know who they are related to? Most likely they do not, but instead follow simple, innate rules: if, for instance, a Belding's ground squirrel only sounds warning calls when lactating or around conspecifics with whom it has had contact since childhood, then a pattern will automatically emerge in which the animals only sound alarms when close relatives were in the vicinity.

Hundreds of studies in the past several decades have demonstrated how animals behave according to Hamilton's theory of kin selection. Belding's studies on ground squirrels are certainly only comprehensible in light of it: since the females are usually surrounded by other female kin, sounding warnings contributes to the survival of their sisters and nieces, thus increasing both their indirect and overall fitness. In contrast, a female whose female kin are no longer alive does not increase her indirect fitness by warning conspecifics of predators – she only increases her own chances of being eaten, and thus reduces her own ability to have children in the future. The same is true for males, who are generally surrounded by non-relatives. It is thus understandable why males and females without living kin in their vicinity take on the risk of sounding alarm calls very rarely or never at all.

Kin selection can also explain communal suckling. In their natural habitat, female mice and lions do not feed just any young alongside their offspring: typically they only feed the offspring of their close female relatives, especially their sisters. Once again, we can see how seemingly altruistic behaviour actually serves to increase an individual's overall fitness.

For the same reason, offspring in many birds and mammal species do not leave their natal group after they become sexually mature, but instead stay with their parents to help raise their younger siblings. We can see the positive effect of such behaviour in species like the African black-backed jackal, whose number of surviving young increases significantly with the number of helpers providing the mother and newborns with food and protection. Along with basic cub-rearing experience, those who stay behind gain an indirect fitness advantage by helping their siblings survive.

One of the most extreme cases of seemingly altruistic behaviour is the aforementioned formation of sterile castes. Ants, bees, and wasp societies in particular all include groups that forgo reproduction in service of the state. All these groups, Hamilton noted, belong to the Hymenoptera, a biological group characterised by one genetic particularity: haplodiploidy.

What does this mean? In haplodiploid species, females, who develop from fertilised eggs, have a double set of chromosomes known as diploids. Each of their genes consists of two alleles; one from the father and one from the mother. Males, in contrast, develop from unfertilised eggs. Accordingly, they have no father and thus have a haploid, or single set of chromosomes, in which each gene consists of only one allele.

Hamilton noticed that haplodiploidy (haploid males, diploid females) in a population has surprising effects on the degree of relatedness between animals. In contrast to all diploid species, sisters are more closely related to one another than mothers are to their daughters: the daughter has a 50 per cent probability of inheriting a particular allele from her mother, while the probability of two sisters sharing an identical allele is 75 per cent due to their father's single set of chromosomes.

Ant, bee, and wasp societies consist exclusively of females, and the sterile castes of such states are comprised of sisters who help their mother, the queen, to raise other sisters. These insects do not add to their direct fitness by forgoing reproduction, but they acquire a considerable degree of indirect fitness by helping their mother to produce further sisters to whom they are very closely related. In fact, in terms of pure arithmetic, sterile individuals do more to maximise their overall fitness by helping to raise their sisters than they would by bearing their own offspring. Hamilton was thus able to demonstrate that the evolution of extreme forms of altruism between closely related individuals is promoted by haplodiploidy, and in doing so answered one of the great unresolved questions of Darwin's theory of evolution. When seemingly altruistic behaviour occurs in the animal kingdom, Hamilton's theory of kin selection explains why most of it is directed towards closely related conspecifics.

AID BETWEEN NON-RELATIVES

While kin selection explains much of what appears to be altruistic behaviour in the animal kingdom, there are also cases of unrelated conspecifics helping one another. Common vampire bats, for example, live in large groups in Central and South America, where they roost in caves or hollow trees and often engage in a much-cited and curious behaviour: blood sharing. As the name suggests, these animals feed on the blood of larger mammals like cattle or horses whom they bite during nocturnal forays. They need to feed constantly: without fresh blood, they will die after only a few days. A study in Costa Rica showed that vampire bats tend to help their group members in a roost who are unable to obtain enough blood. As predicted by kin selection theory, females preferentially shared blood with their own offspring and other close relatives. Remarkably, however, they also shared with unrelated group members in need, but only those with whom they had a preexisting close relationship and could reasonably count on the favour being repaid.

Blood sharing is therefore not random, but also not fully explained by kinship. Rather, it seems to be guided at least in part by the principle that the American evolutionary biologist Robert Trivers has called 'reciprocal altruism': If I help you, you help me. Bats will only share blood with others who would be willing to share themselves.

Another often cited example of reciprocal altruism among non-relatives can be found in Anubis baboons. In many groups, an inferior male might have a chance of reproducing with a female if another inferior helps him distract the alpha, who is normally the only one who gets to mate. Interestingly, if one inferior distracts the alpha for another, then the second inferior will help the first at the next opportunity.

Trivers formulated his theory of reciprocal altruism about 50 years ago, although the term has fallen out of use in recent years: animals who offer help to others can also expect it in return, and so their behaviour is not truly altruistic. Instead, we talk about reciprocity – that is, time-based mutual aid – which occurs more frequently in the everyday life of many animals than originally assumed. Many species of monkey, for instance, form habits based on reciprocity, like sharing their food in the afternoon with conspecifics who will scratch their backs in the morning.

Non-relatives cooperate most frequently when all individuals can reap immediate benefits. African wild dogs, for example, hunt gazelles and antelopes in packs. The more dogs in the pack, the more successful the hunt in both overall and individual terms. Thus, each dog benefits individually from helping each other.

The evolution of mutual aid between non-relatives is simple to explain: there are immediate advantages to helping conspecifics, be they better access to food, better protection from enemies, or other factors that contribute to survival and are thus also vital to passing on one's own genes.

Ultimately, detailed behavioural analyses prove cooperating with non-relatives to be just as self-serving as helping one's kin as there are hardly any forms of cooperation that do not benefit the helper as well as the helped.

KILLING CONSPECIFICS TO GAIN REPRODUCTIVE ADVANTAGE

The current state of research tells us that animals are 'programmed' by natural selection to pass on their own genes with maximum efficiency, and accordingly behave in line with genetic self-interest rather than the good of the species. If helping conspecifics benefits an animal's overall fitness, then an animal will behave seemingly altruistically. But if it can better contribute to its fitness in other ways, it will: animals will indeed threaten, fight, coerce, deceive, or even kill their conspecifics.

Thus, males in species from ungulates to primates to dolphins will sexually harass their female counterparts, and younger and lower-ranking males in a wide variety of animal societies will force females to copulate against their will. These forms of coercion cause significant harm to the females, but they increase the males' chances of passing on their own genes.

Another striking example of this kind of brutality is infanticide, which can often be observed in stable groups of long-living mammals when foreign males invade. In New- and Old-World monkeys (including great apes), rodents, and carnivores foreign males will regularly kill the off-spring of their conspecifics, so much so that infanticide accounts for a significant proportion of juvenile mortality. In mountain gorillas, for example, more than one-third of juvenile mortality is due to the killing of infants by foreign males.

This practice is particularly well documented in African lions. A lion pride typically consists of several related females and two to three unrelated males. While the females are born in the pride and stay there for the rest of their lives, the males, who are often brothers, remain part of the group for only about two years once they reach their prime. They use this time to reproduce with the females but are also repeatedly exposed to attacks by foreign males. At first the males can defend the pride, but ultimately, they are defeated by younger and stronger rivals who take over for the next two years until they are driven away themselves.

Remarkably, such power shifts are often accompanied by infanticide, as the new heads of the pride bite the unweaned young of their predecessors to death. In the past, this was seen by some biologists as accidental or disturbed social behaviour, while others even interpreted it as a service to the species, allowing the survivors greater access to resources. But if we view infanticide from the perspective of self-interest we arrive at a very different explanation.

Males only have a good chance of reproducing during the two years or so that they are members of a pride. Meanwhile, as long as a female is nursing, she does not ovulate – a hormonal process also known in humans as lactational amenorrhoea – meaning she cannot become pregnant again until her young are weaned. If the new males kill the unweaned young after taking over the pride, then the females will be ready to reproduce much sooner and will in turn increase the potential reproductive success of the males. Through this lens, infanticide is neither a behavioural disorder nor an act in the service of the species, but a way of enabling the new males in a pride to maximise their own reproductive success.

But what about the females? The killing of their offspring is an obvious blow to their reproductive success, and indeed, they often try to prevent it. First, they avoid new males, and, if that does not work, they threaten or attack them, or even leave the pack with their unweaned offspring. But these countermeasures are not often successful, as the males are typically stronger than the females and young are unlikely to survive outside of the pride. Infanticide in fact is an example of one of the ways in which the self-interest between males and females in a species can be at odds, and thus certain conflicts are preordained to happen.

Conflicts – sometimes fatal conflicts – can also occur between siblings. Mothers having or being willing to give less milk than their offspring demand often leads to competition. In pigs, for example, fierce disputes lead to the formation of a 'teat order' in the very early days of lactation: the strongest piglets suckle from the front nipples, which provide more and higher-quality milk, while the weaker piglets have to make do with the nipples at the rear. It is no surprise, then, that the piglets who feed from the front develop significantly better than those at the back.

One of the most extreme examples of the dramatic consequences of sibling rivalry comes from the spotted hyenas who inhabit much of Africa. Female spotted hyenas usually give birth to two young, whom they suckle for a good year. At birth, the babies' canines and incisors are already well developed, allowing the siblings to aggressively compete for dominance from the very first days of life. The outcome of these disputes has serious developmental consequences as the dominant sibling gets access to more milk. The more effort a mother has to make to find food in times of scarcity, the more reduced her milk production, and thus the more exclusive a resource it becomes, causing siblings to fight more intensely, often to the death. When one sibling is killed, the other rapidly gains weight and has a significantly better chance of survival.

According to kin selection theory, we generally expect siblings to help rather than hurt each other. But under difficult ecological conditions, when killing a brother or a sister might increase the chances of one's own survival, siblicide will indeed occur. Like infanticide, the killing of siblings is neither a sign of disturbed behaviour nor an act to benefit the species as a whole: it is a means of serving an individual's self-interest.

A number of studies have also shown that fights between adult males lead to serious injuries and deaths much more frequently than originally thought. In male red deer, for example, 20 to 30 per cent incur permanent injuries from fighting over their lifetime. East Asian water deer can inflict fatal wounds with their canines, while many other deer and antelope species can impale opponents with their antlers or horns. Clearly, the dogma from classical ethology that animals do not kill conspecifics is a myth: no such innate aversion exists.

We can see similar examples in the famous 'chimpanzee wars' that Jane Goodall first described over 40 years ago, during which males of a large group of chimpanzees joined forces in the Gombe Stream National Park in Tanzania and killed all of the males in a neighbouring smaller group over the course of a few years. Today we know that Goodall's observation of such warlike aggression was not an isolated case. In total, scientists have documented about 150 cases of chimpanzees killing conspecifics in the last few decades, two-thirds of which were due to violent conflicts between different groups. Attackers almost always outnumbered their victims: in one extreme case, up to 28 males of a group allied and attacked another, killing other males as well as mothers and children. Such attacks have been found to occur during almost all long-term observational studies of chimpanzees.

Why do chimpanzees behave like this in the wild? Because it benefits the perpetrators: by killing rival non-relatives in low-risk situations, they are able to expand their territory and gain access to important resources, food, or mates. There is therefore every reason to believe that 'chimpanzee wars' are a result of natural selection, rather than the popular claim that such clashes are caused by human disturbances to the habitat like deforestation, hunting, or feeding.

MALES AND FEMALES

The sociobiological revolution has also led scientists to reassess gender roles in the animal kingdom. We now see the behaviour of females, for example, in a completely different light. But to understand these changes, we must briefly discuss one of Darwin's key concepts: that animals must not only survive, find food, and evade enemies but also compete for mates. The better they are able to compete, the greater their reproductive success: a process Darwin called sexual selection.

Sexual selection takes two forms: in the first, which is known as intrasexual selection, males usually compete to decide who gets the opportunity to mate – a widespread practice in the animal kingdom. Male red deer, for instance, have what is known as a 'rutting season' during which they become increasingly incompatible. Conflicts start off with loud roaring duels that increase in volume and frequency as they escalate. These are often sufficient tests of strength in themselves, as males who can endure them usually also have success in physical clashes. But if the deer can declare no clear winner this way, they move on to the next stage: 'parallel walking', during which the two rivals strut up and down a short distance and try to estimate each other's strength until one of them retreats. If neither does, the deer finally fight. No matter how the winner is determined, the result is the same: the winner gets to mate with the female.

Darwin's idea of intrasexual selection quickly gained acceptance in the scientific community. While this view is in many cases correct, for over a hundred years it was linked with the belief that females are passive and cautious in response to male sexual advances. Indeed, until the 1980s, established textbooks taught that natural selection demonstrated a strong preference for 'female prudery'.

Yet Darwin had by no means considered females passive. Rather, he assumed that a second form of sexual selection existed alongside the male competition for females: a process he termed intersexual selection, in which females decide whom to have as a mate. The Victorian Zeitgeist likely caused both the public and the scientific community to ignore this idea for more than a century. Now, however, we know of countless examples from the animal kingdom in which the females actually determine which males get to reproduce and are by no means passive recipients of male courtship and sexual behaviour. In many species, in fact, almost all sexual interactions are initiated by the females.

Genetic fingerprinting has also significantly contributed to the recent change in our understanding of female sexual behaviour. Since the 1990s, it has been possible to determine paternity from a few skin cells, hairs, or feathers with a high degree of certainty.

Studies on songbirds in particular have yielded huge surprises. Songbirds have forever been considered the very symbol of fidelity: as the story goes, a male and female form a pair, build a nest, protect their eggs, and raise their offspring together. But genetic fingerprinting has revealed that a high proportion of the offspring in a pair's nest do not come from the males who feed and care for them. Studies on tits have shown that up to 80 per cent of nestlings are the products of 'extramarital' affairs with neighbouring males. At first scientists speculated that the males were initiating these encounters, but it has since become clear that females are the ones seeking them out. Why?

Infanticide among lions has already shown us how males and females can have quite different interests. Both sexes try to pass on their own genes with maximum efficiency, inevitably resulting in conflict. So let us consider the situation from the female songbird's perspective: to achieve the highest possible reproductive success, she should mate with the highest-quality male and live in the most resource-rich environment. But both options at once are not available to most females: often the best territories are already occupied by other females who vehemently defend them. If her own mate is not up to par, it benefits a female to find a higher-quality male from an adjacent nest to fertilise her eggs.

For their part, the males do their utmost not to be deceived into raising the young of their rivals. Males of many species tend to guard their mates especially vigilantly during the period in which their eggs are fertilised, sometimes resorting to perplexing extremes. When male barn swallows do not find their mates in their nests, for example, they sound alarm calls normally reserved for predators. Upon hearing such a call, all swallows in the vicinity will immediately stop what they are doing – including mating with another partner.

ON THE SOCIOBIOLOGY OF WILD GUINEA PIGS

Some of our own research has also contributed to the changing view of female behaviour. Our comparison of different species of wild guinea pigs in particular has significantly contributed to a reimagining of the role of the female and the evolution of mating and social systems as a whole.

As far as we know, there are about 15 different species of wild guinea pig living in South America today. Wild guinea pigs are grouped with maras and capybaras in the family *Caviidae* – known colloquially as guinea pigs – to which domestic guinea pigs also belong. The common wild cavy is the ancestral form of domestic guinea pigs and is particularly well studied, as is a second species, the common yellow-toothed cavy. Common yellow-toothed cavies differ markedly from common wild cavies in their behaviour and appearance, however, and as a result the two species belong to different genera and cannot be interbred.

We have studied both species at our institute for many years, and have observed them in their natural habitat in South America. Yellow-toothed cavies can usually be kept in enclosures with many other females and males relatively stress-free. All the animals in such an environment will usually huddle together, nestling closely above, below, and next to each other. There are no particular bonds or friendships among different group members: every yellow-toothed cavy interacts with every other, sometimes in a friendly way, sometimes in an aggressive one. No matter what, all the animals will reconvene in a large pile when it is time to huddle again.

One particular observation led us to completely re-evaluate the female sexual behaviour of yellow-toothed cavies. One day, a female in the group began to zig-zag around the enclosure, calling out loudly to all the surrounding males. They all ran after her, and the highest-ranking of the group tried to capture her attention but could not keep track of her unpredictable path. Suddenly the female stopped, and the male right behind her ran up and mated with her. Then she continued running again, and the males followed, until she once again stopped and allowed the next male in line to mount her. The chase continued like this until all the males had copulated with the female. The female, it certainly seemed, was the one initiating mating behaviour.

We were able to prove that this was indeed the case using what is called a preference test. First, we built an apparatus consisting of a central enclosure connected by hallways to four outer chambers. Then we placed the female in the central enclosure and a male in each of the chambers, allowing the female free access to any of the outer rooms but restricting the males to the rooms they occupied. Thus, the female was free to choose her mate, and in fact most of the females we observed did run from one enclosure to the next and copulated several times with many of the males.

Such behaviour almost always leads to offspring from several different fathers, as we were able to show in follow-up studies using genetic fingerprinting. Young born in the same litter are thus usually the progeny of several different males. This is not only true for yellow-toothed cavies in captivity, but also in the wild: studies of yellow-toothed cavies in their natural habitat in Argentina have also proven that 50 to 80 per cent of litters have multiple paternities.

Why do females go out of their way to mate with more than one male? In sociobiological terms, such behaviour is logical if it leads to higher overall reproductive success. This is indeed the case: in a now-famous experiment, we paired several female yellow-toothed cavies with either one or four different males. Regardless of the situation, the females almost always became pregnant, and the number of offspring they gave birth to did not differ significantly. But the newborns from mothers who had mated with multiple partners showed almost twice the survival rate as the young from the mothers who had mated with only one. This was the first experiment to show how mating with multiple partners increases mammalian females' reproductive success.

Why is this? Matthias Asher, a researcher on our team, devoted his doctoral thesis to finding the answer. Asher studied common yellow-toothed cavies in their sparsely vegetated, semi-desert natural habitat in Argentina. The guinea pigs living there must roam fairly far to find food, meaning males cannot monopolise any number of females and no social bonds tend to form between individuals at all. Meanwhile, a large number of birds of prey, snakes, and other predators are always lurking; the animals only have a chance of surviving in areas where thorn bushes offer places to hide. These areas are often very far from each other: the guinea pigs do not often risk moving between them. Thus, they tend to stay in the places they were born, which consequently results in high degrees of inbreeding. This is problematic: as we were able to demonstrate with the help of colleagues specialising in reproductive medicine, the quality of the males' sperm declines as a result.

How should a female in such an environment behave? Of course she should seek out a partner with high-quality sperm. But how can she tell which males are genetically superior? She cannot – so letting the sperm of several males compete to fertilise her eggs is the best course of action. Indeed, in common yellow-toothed cavies, promiscuous mating behaviour leads to sperm competition. As a result, sperm of low quality is excluded from fertilisation. This is the most likely explanation for why the young of females who have mated with multiple partners tend to survive at higher rates than those whose mothers have mated with only one.

Are all mammalian females promiscuous? Not at all – as Asher discovered in another study on the closely related common wild cavy. In their natural habitat in South America, these animals live in wet grasslands where food is almost always plentiful and evenly distributed. The animals there accordingly have much smaller foraging areas than common yellow-toothed cavies; it is even common for several females to graze in the same area. Under these conditions, strong males can easily mark one, two, or three females as their own and successfully defend them against rivals, causing pairs or small harems to form in line with the distribution of females in a certain locale.

But like yellow-toothed cavies, wild cavies also face considerable pressure from predators in their natural habitat. The most suitable environments are thus those that have both open, grassy areas where the animals can feed, and dense vegetation where they can hide in case of danger.

When wild cavies sense a threat, they tend to hide and remain motionless until they believe it is safe to move again. Mammals who use this avoidance strategy generally try not to draw attention to themselves and keep away from conspecifics. Wild cavies are no exception in this respect, which may be part of the reason why they form small groups of a male and several females instead of large ones. Our studies have shown that no single form of social organisation prevails across all guinea pigs, but rather individual species tend to adapt to their ecological niche: a pattern consistent with behavioural ecology theory.

Genetic paternity studies have shown that the two species of wild guinea pig also differ in terms of their mating systems. While the vast majority of litters born to common yellow-toothed cavies show multiple paternities, the young of common wild cavies tend to come from the male who lives in a pair or harem with the mother.

It is reasonable to assume that these differences between species are also related to the behaviour of females. Indeed, when we gave female wild cavies the opportunity to choose one of four mates, they behaved quite differently from yellow-toothed cavies: they looked at all the males closely before choosing one social partner who later became the father of their offspring. Thus, the differences between reproductive and social systems across species may be partially driven by the ways in which females choose their mates.

About twenty years ago we had to refresh our population of common yellow-toothed cavies in the lab to avoid issues connected with inbreeding. The animals we introduced from Bolivia differed slightly from those in our group in physique and colour, variations we at first explained by the animals' different regions of origin in South America. But the new guinea pigs proved to have no success in mating with our group, so we took a closer look at them with colleagues from the Senckenberg Museum in Frankfurt. Our findings caused a small sensation: the new arrivals turned out not to be common yellow-toothed cavies at all, but a new, previously unknown species. Since we discovered them in Münster, we named them *Galea monasteriensis*: the Münster yellow-toothed cavy. For a while this was the most newly discovered mammal species on the planet.

No studies have so far been conducted on the Münster yellow-toothed cavy in its natural habitat. The research from our institute suggests that it is a monogamous species, as males tend to be incompatible with other males, and females likewise with other females. When the right male and female meet, however, they get along immediately, quickly form a harmonious pair, and reproduce. Studies on other monogamous mammals have shown these tendencies to be typical. The Münster yellow-toothed cavy has demonstrated other characteristics of monogamous species as well: when we gave females the choice between different partners, they inspected all potential mates before choosing one with whom to form a strong social bond and reproduce.

So, as luck would have it, we were able to simultaneously study the social systems of three different species of guinea pig at once: common yellow-toothed cavies who did not form any strong social bonds with conspecifics, common wild cavies who formed small harems, and Münster yellow-toothed cavies who lived in pairs. Such a set-up is ideal for systematically testing certain sociobiological theories. Paternal behaviour towards offspring, for example, should differ markedly among these species, as there are significant differences in how likely a given male is to be the father of any offspring: it is somewhat unlikely for common yellow-toothed cavies, probable for common wild cavies, and certain for Münster yellow-toothed cavies. For the purposes of maximising fitness, it only benefits males to invest time and energy in offspring who are surely their own.

Oliver Adrian, a researcher from my team, tested this hypothesis by analysing the paternal behaviour of all three species. The males of the Münster yellow-toothed cavies cared intensively for their offspring, playing with them frequently. The males of the common yellow-toothed cavy, in contrast, were uninterested and occasionally aggressive, while the common wild cavy males behaved somewhere in between: they were not aggressive towards their young, but played with them significantly less than the Münster yellow-toothed cavy males did.

Overall, Adrian found that the more sure a father could be of his paternity, the more he invested in his offspring. This result impressively confirms what sociobiological theory has predicted: in terms of paternal care, too, animals do not behave to benefit the species as a whole, but to maximise their own reproductive success.

CONCLUSION

Numerous studies have shown that animals do not behave for the good of the species but are programmed by natural selection to pass on copies of their own genes. If they best achieve this goal by cooperating with conspecifics, then they do – but if they can better achieve it by means of coercion, aggression, or killing members of their own species, they will do so as well. Behaving with the goal of maximising one's own fitness inevitably leads to conflict: between males, between females, between siblings, and between the two sexes. Among other things, this perspective has revolutionised the scientific understanding of female roles and behaviour. Far from being passive recipients of males' actions and advances, they behave actively and efficiently to maximise their own fitness.

CHAPTER 8

Animals Like Us

A Summary

BEHAVIOURAL BIOLOGY HAS FUNDAMENTALLY CHANGED OUR scientific understanding of animals. We now doubtlessly better understand their thinking, feelings, and behaviour, and in the process are discovering how similar we really are to each other, indeed to a far greater extent than we could have even a few years ago imagined.

Admittedly, the degree to which this is true is not the same across all animals: we are clearly closer to chimpanzees, dolphins, dogs, or mice than we are to ants, starfish, snails, or amoebas. We are vertebrates, and we are mammals, and we share a comparable brain, hormones, and nervous system with all other creatures in this class. Thought, emotions, and behaviour are all ultimately functions of these systems. The more comparable living beings are in this respect, the more similar they are to each other.

Today we know that mammals are certainly not automatons who react reflexively and rigidly to external stimuli. They do not develop along a predetermined path; rather, as with us, environmental, learning, and social experiences play a major role in shaping their behaviour. Even prenatal influences can profoundly alter their brain and behavioural development, while experiences in early childhood, a phase during which an organism's nervous system is particularly sensitive, can have a particularly lasting effect. But later phases are also important: in social species, for instance, animals acquire essential interaction skills during adolescence. Although learning is of particular importance during these early stages, behaviour is plastic well into old age. Like humans, other mammals are 'open systems', remaining receptive to new influences and experiences throughout their lives.

And, as with humans, mammalian behaviour is controlled by a number of factors. The way in which a particular behaviour is triggered and managed usually depends on a series of internal and external features, like sex, age, social status, experience, hormonal state, and genetic predisposition; it is neither possible nor appropriate to attribute a certain behaviour to any single one. Hormones alone, for example, are insufficient to explain aggression. Researchers have identified numerous genes in recent years that are involved in the control of behaviour, but they are not the sole determinants of it at all. In principle this is unsurprising, as behaviour always arises from the complex interaction between genes and environment. As we have discussed, genetically 'unintelligent' mice who are raised in a learning-conducive environment perform better on learning tests than genetically 'intelligent' conspecifics who have grown up with minimal stimulation. On the other hand, even the tiniest genetic variations can cause individuals to react quite differently to the same situation. Acquired behavioural traits, moreover, can be passed on through epigenetic inheritance. In all these aspects, humans and other mammals do not appear to differ.

We know well that the dynamic between genome and environment in the early developmental stages leads humans to develop unique personalities. But our non-human relatives also develop distinctive characteristics in the course of behavioural ontogeny: no two chimpanzees are alike; every mouse, every tit is different from its conspecifics in some way. The discovery of stable, long-term 'animal personalities' has also brought the individual into focus, so much so that individuality is considered a basic feature of behaviour.

Another great feature of existence we share with our non-human relatives is the relationship between our environment, behaviour, and stress. The patterns we see in our mammalian relatives apply almost identically to humans: when individuals are integrated into a stable social system with clear relationships, they hardly experience stress reactions. Social instability, in contrast, causes a strong release of stress hormones that in the long term can lead to an increased susceptibility to disease. The most effective buffers against stress are the same for humans and animals: good social bonding partners. The better the social relationships, the greater they protect against stress.

Like us, other animals have emotions, and experience both positive and negative feelings to a degree that depends on their situation and personality. According to the current state of research, their basal emotions like fear, anxiety, and joy are controlled by the same set of neural circuitries as ours are. Of course, scientists have yet to answer the question of the breadth and type of emotion that animals can experience: there is good reason to assume that not all human emotions are present in animals, nor all animal emotions in humans. Overall, however, the basic finding remains that non-human mammals are sentient creatures with emotions that are in large part comparable to ours.

The scientific community has made great strides in accepting the humanity in animals, although we often ignore how this finding can work the other way as well. If behavioural development is an open-ended process in other mammals, then we should not assume it is fixed at conception, birth, or even the end of childhood in humans either. If other mammals can learn how to interact peacefully with strangers during adolescence, then there are no biological reasons why humans should not also be able to do this. If genetics do not determine behaviour in our relatives, then they should not in us. Social relationships and positive emotions should be the best cure for stress and illness in humans and our animal relatives alike.

The findings from behavioural research that have most stoked the public's interest in recent years have been in the field of cognitive performance. Such findings, it seems, challenge our self-image as humans as we have long considered ourselves to be the sole proprietors of rationality. But today we know that certain animals cannot only learn but also think: they can craft tools, learn how to use them, and even pass on their innovations through generations. Some animals can recognise themselves in a mirror; others can predict what their conspecifics are able to perceive and use this knowledge to their own ends. Taken together, these findings may well indicate that animals like apes, elephants, or dolphins have self-awareness: like humans, they know who they are.

Humans doubtlessly possess the highest cognitive abilities of all living creatures. But as countless studies on birds have shown, evolution indeed leads to higher cognitive function, but not in a direct line to our species. Scientists long believed that great apes, our closest relatives, were the most 'intelligent' animals, although research now suggests that corvids may cognitively be on the same plane. Because the lineages of mammals and birds split hundreds of millions of years ago, the evolution of higher cognitive function is by no means a straight path that leads only to humans at the pinnacle: in other words, humans are indeed not the 'pride of creation'.

What, then, ultimately distinguishes humans from animals? Many species can communicate with a high degree of efficiency and sophistication using phonetic utterances, and recent research has started to frame animal language as closer to our own. But the complex ways in which humans communicate – how we exchange information about the past, present, and future, for example – is not known to other species.

While animals can think and perhaps even possess self-awareness, they seem to reflect on themselves and the world very little, if at all. They can plan ahead for a few hours or days, for instance, but cannot consciously project weeks, months, or years into the future. They can teach their offspring how to use certain tools or hunt certain prey, but they can only educate them with certain norm-guided goals to a rudimentary degree at best. Animals can come up with certain innovations that spread across their communities and are passed down through generations, but these innovations are not further developed or improved upon by others. As the American developmental psychologist Michael Tomasello aptly summed it up, animals possess no significant cumulative cultural evolution.

Scientists often debate whether the differences between humans and animals are categorical or exist on a spectrum. On the one hand, an animal has never composed a symphony, written a novel, built a cathedral, or formulated a plan of action to combat climate change. On the other, animals are capable of comparable sorts of cognitive feats to two-, three-, or four-year-old children; there is hardly a human characteristic or ability that is not present, at least in basic form, in our animal relatives. From a behavioural point of view, there is no doubt that animals have moved closer to us. Whether the remaining difference is quantitative or qualitative cannot, however, be decided on the basis of behavioural data. Ultimately, it is up to the individual to decide.

But public debate often forgets another key way in which animals share a good deal of human nature. Human concepts of morality tend to push a certain image of animals as inherently good. True, the social life of many mammals is characterised by prosocial behaviour, extensive cooperation, and harmony. As the Dutch-American primatologist Frans de Waal discovered in several impressive studies, individuals from some species even have a sense of fairness, can understand and share the emotions of others, comfort group members, and possess sophisticated abilities to resolve conflicts.

But it is also true that these same animals will threaten, fight, coerce, rape, and kill their conspecifics if it serves their own interests. Even warlike conflicts can emerge among groups of chimpanzees. These findings stand in stark contrast to the classical ethological dogma that animals behave for the good of their species – in reality, a general aversion towards killing conspecifics appears not to exist.

Today we have a much more nuanced view on animal behaviour than previously existed. Instead of behaving for the good of their species, animals behave according to the principle of self-interest, acting to pass on their own genes with maximum efficiency. If they can best do this through cooperation, then they will, but by that same logic they will also not shy away from brutalising one another.

In the end, animals are not the ones who are 'like us, but better' – rather, humans are the only species who might be able to shake the grip of self-interest through the capacities of our culture and intellect: human rights, education towards peace and tolerance, and equality before the law.

CONCLUSION

Behavioural biology has undergone several paradigm shifts in recent decades. Scientists have changed their perspectives from group selection to individual – and kin – selection, from instinct to interactions between genes and environment and epigenetic inheritance, from developmental rigidity to lifelong plasticity, from templated behaviour to animal personalities, and from conditioning to cognition. They have moved from neglecting animal emotions to embracing them, and from pathologising behaviour to understanding its adaptive advantages; they have realised that a humane life in captivity means more than physical health and fertility. These shifts have revolutionised the way we see animals, helping us to better understand and provide them with a more humane life. At the same time, these shifts have demonstrated again and again how similar we humans and animals really are to each other: they are starting to close in the gap between us. The more we investigate, and the closer we look, the more we see the humanity in our fellow creatures.

Bibliography

CHAPTER 1

Darwin, C., *The Expression of the Emotions in Man and Animals.* (Reprint) The University of Chicago Press, Chicago, London, 1965 (Original: 1872).

Darwin, C., *Die Entstehung der Arten.* Neudruck Reclam, Stuttgart, 1981 (Original 1859).

Franck, D., *Eine Wissenschaft im Aufbruch. Chronik der Ethologischen Gesellschaft 1949–2000.* Verlag Niel & More, Hamburg, 2008.

Frisch, K. v., *Tanzsprache und Orientierung der Bienen.* Springer, Berlin, Heidelberg, 1965.

Immelmann, K., *Wörterbuch der Verhaltensforschung.* Verlag Paul Parey, Berlin, Hamburg, 1982.

Kaiser Friedrich der Zweite, *Über die Kunst, mit Vögeln zu jagen.* Insel-Verlag, Frankfurt, 1965.

Lorenz, K., Der Kumpan in der Umwelt des Vogels. *Journal für Ornithologie* 83: 137–213 and 289–413, 1935.

Lorenz, K., Vergleichende Bewegungsstudien an Anatiden. *Journal für Ornithologie* 89: Ergänzungsband, 194–293, 1941.

Naguib, M., *Methoden der Verhaltensbiologie.* Springer-Verlag, Berlin, Heidelberg, 2006.

Pfungst, O., *Der kluge Hans (Nachdruck der Originalausgabe von 1907).* Fachbuchhandlung für Psychologie, Frankfurt, 1983.

Tinbergen, N., *The Study of Instinct.* Oxford University Press, London, 1951.

Tinbergen, N., On the Aims and Methods of Ethology. *Zeitschrift für Tierpsychologie* 20: 410–433, 1963.

Zippelius, H. M., *Die vermessene Theorie.* Friedr. Vieweg & Sohn Verlagsgesellschaft, Braunschweig, Wiesbaden, 1992.

CHAPTER 2

Bradley, A. J., McDonald, I. R., Lee, A. K., Stress and Mortality in a Small Marsupial (*Antechinus stuartii*, Macleay). *General and Comparative Endocrinology* 40: 188–200, 1980.

Cannon, W. B., *Bodily Changes in Pain, Hunger, Fear and Rage*. Branford, Boston, 1929.

Christian, J. J., Phenomena Associated with Population Density. *Proceedings of the National Academy of Sciences of the USA* 47: 428–449, 1961.

Gesquiere, L. R., Learn, N. H., Simao, M. C. M., et al., Life at the Top: Rank and Stress in Wild Male Baboons. *Science* 333: 357–360, 2011.

Hennessy, M. B., Kaiser, S., Sachser, N., Social Buffering of the Stress Response: Diversity, Mechanisms, and Functions. *Frontiers in Neuroendocrinology* 30: 470–482, 2009.

Henry, J. P., Stephens, P. M., *Stress, Health, and the Social Environment. A Sociobiologic Approach to Medicine*. Springer, New York, 1977.

Kaplan, J. R., Manuck, S. B., Clarkson, T. B., Lusso, F. M., Taub, D. M., Social Status, Environment, and Atherosclerosis in Cynomolgus Monkeys. *Arteriosclerosis* 2: 359–368, 1982.

Koolhaas, J. M., Korte, J. M., de Boer, S. F., et al., Coping Styles in Animals: Current Status in Behaviour and Stress-Physiology. *Neuroscience & Biobehavioral Reviews* 23: 925–935, 1999.

McEwen, B. S., Wingfield, J. C., The Concept of Allostasis in Biology and Biomedicine. *Hormones and Behavior* 43: 2–15, 2003.

Sachser, N., Dürschlag, M., Hirzel, D., Social Relationships and the Management of Stress. *Psychoneuroendocrinology* 23: 891–904, 1998.

Sachser, N., Kaiser, S., Meerschweinchen als Sozialstrategen. *Spektrum der Wissenschaft* January 2010: 56–63, 2010.

Selye, H., *Stress*. Acta, Montreal, 1950.

Von Holst, D., The Concept of Stress and Its Relevance for Animal Behaviour. *Advances in the Study of Behavior* 17: 1–131, 1998.

Young, C., Majolo, B., Heistermann, M., Schülke, O., Ostner, J., Responses to Social and Environmental Stress Are Attenuated by Strong Male Bonds in Wild Macaques. *Proceedings of the National Academy of Sciences of the USA* 111: 18195–18200, 2014.

CHAPTER 3

Broom, D. M., Johnson, K. G., *Stress and Animal Welfare*. London, 1993.

Clubb, R., Mason, G., Animal Welfare: Captivity Effects on Wide-Ranging Carnivores. *Nature* 425: 473–474, 2003.

Current Biology, Biology of Fun, 25th Anniversary Special Issue, 1, R1–R30, 2015.

Dawkins, M. S., From an Animal's Point of View: Motivation, Fitness, and Animal Welfare. *Behavioural and Brain Sciences* 13: 1–9 and 54–61, 1990.

Harris, C. R., Prouvost, C., Jealousy in Dogs. *PLoS One* 9: 7, e94597, 2014.

Kaiser, S., Classen, D., Sachser, N., Auswirkungen unterschiedlicher struktureller Anreicherungen auf das Spontanverhalten weiblicher Labormäuse (Stamm NMRI). In *Aktuelle Arbeiten zur artgemäßen Tierhaltung 1998*. KTBL-Schrift 382: 56–62, 1998.

Panksepp, J., Beyond a Joke: From Animal Laughter to Human Joy? *Science* 308: 62–63, 2005.

Paul, E. S., Harding, E. J., Mendl, M., Measuring Emotional Processes in Animals: The Utility of a Cognitive Approach. *Neuroscience & Biobehavioral Reviews* 29: 469–491, 2005.

Richter, S. H., Sachser, N., Kaiser, S., Tiere und Emotionen. In *Handbuch Tierethik*, Ach, J. S., Borchers, D. (Eds.), J. B. Metzler, Stuttgart, 2018.

Sachser, N., *Sozialphysiologische Untersuchungen an Hausmeerschweinchen. Gruppenstrukturen, soziale Situation und Endokrinium, Wohlergehen.* Verlag Paul Parey, Berlin and Hamburg, 1994.

Sachser, N., Was bringen Präferenztests? In *Aktuelle Arbeiten zur artgemäßen Tierhaltung 1997* KTBL Schrift 380, Darmstadt, 9–20, 1998.

Sachser, N., What Is Important to Achieve Good Welfare in Animals? In *Coping with Challenge: Welfare in Animals Including Humans*, Broom, D. M. (Ed.), Dahlem Workshop Report 87, Dahlem University Press, Berlin, 31–48, 2001.

Sachser, N., Neugier, Spiel und Lernen: Verhaltensbiologische Anmerkungen zur Kindheit. *Zeitschrift für Pädagogik*, 475–486, 2004.

Sachser, N., Richter, S. H., Kaiser, S., Artgerecht – tiergerecht: eine biologische Perspektive. In *Handbuch Tierethik*, Ach, J. S., Borchers, D. (Eds.), J. B. Metzler, Stuttgart, 2018.

Schmidt, C., Sachser, N., Auswirkungen unterschiedlicher Futterverteilungen auf Verhalten und Speichel-Streßhormonkonzentrationen von Breitmaulnashörnern im Allwetterzoo Münster. In *Aktuelle Arbeiten zur artgemäßen Tierhaltung 1996.* KTBL-Schrift 376: 188–198, 1997.

CHAPTER 4

Ambrée, O., Leimer, U., Herring, A., et al., Reduction of Amyloid Angiopathy and Aβ Plaque Burden after Enriched Housing of TgCRND8 Mice. *The American Journal of Pathology* 169: 544–552, 2006.

Belsky, J., Jonassaint, C., Pluess, M., et al., Vulnerability genes or plasticity genes? *Molecular Psychiatry* 14: 746–754, 2009.

Brunner, H. G., Nelen, M., Breakefield, X. O., Ropers, H. H., van Ost, B. A., Abnormal Behavior Associated with a Point Mutation in the Structural Gene for Monoamine Oxidase A. *Science* 262: 578–580, 1993.

Cases, O., Seif, I., Grimsby, J., et al., Aggressive Behaviour and Altered Amounts of Brain Serotonin and Norepinephrine in Mice Lacking MAOA. *Science* 268: 1763–1766, 1995.

Caspi, A., Sugden, K., Moffitt, T. E., et al., Influence of Life Stress on Depression: Moderation by a Polymorphism in the 5-HTT Gene. *Science* 301: 386–389, 2003.

Cooper, R. M., Zubek, J. P., Effects of Enriched and Restricted Early Environments on the Learning Ability of Bright and Dull Rats. *Canadian Journal of Psychology* 12: 159–164, 1958.

Dias, B. G., Ressler, K. J., Parental Olfactory Experience Influences Behaviour and Neural Structure in Subsequent Generations. *Nature Neuroscience* 17: 89–96, 2014.

Epstein, R., Lanza, R. P., Skinner, B. F., Symbolic Communication Between Two Pigeons (*Columba livia domestica*). *Science* 207: 543–545, 1980.

Glocker, M. L., Langleben, D. D., Ruparel, K., et al., Baby Schema in Infant Faces Induces Cuteness Perception and Motivation for Caretaking in Adults. *Ethology* 115: 257–263, 2009.

Glocker, M. L., Langleben, D. D., Ruparel, K., et al., Baby Schema Modulates the Brain Reward System in Nulliparous Women. *Proceedings of the National Academy of Sciences of the USA* 106: 9115–9119, 2009.

Heiming, R. S., Jansen, F., Lewejohann, L., et al., Living in a Dangerous World: The Shaping of Behavioural Profile by Early Environment and 5-HTT Genotype. *Frontiers in Behavioural Neuroscience* 3: 26, 2009.

Immelmann, K., Pröve, E., Sossinka, R., *Einführung in die Verhaltensforschung 4.* Aufl., Wien, 1996.

Lewejohann, L., Reefmann, N., Widmann, P., et al., Transgenic Alzheimer Mice in a Semi Naturalistic Environment: More Plaques, Yet Not Compromised in Daily Life. *Behavioural Brain Research* 201: 99–102, 2009.

Meaney, J. M., Maternal Care, Gene Expression, and the Transmission of Individual Differences in Stress Reactivity Across Generations. *Annual Review of Neuroscience* 24: 1161–1192, 2001.

Sachser, N., Lesch, K. P., Das Zusammenspiel von Genotyp und Umwelt bei der Entwicklung von Furcht und Angst. *Neuroforum* 3: 104–109, 2013.

Seyfarth, R. M., Cheney, D. L., Wie Affen sich verstehen. *Spektrum der Wissenschaft* 2: 88–95, 1993.

Weaver, I. C. G., Cervoni, N., Champagne, F. A., et al., Epigenetic Programming by Maternal Behavior. *Nature Neuroscience* 7: 847–854, 2004.

CHAPTER 5

Brosnan, S. F., de Waal, F. B. M., Monkeys Reject Unequal Pay. *Nature* 425: 297–299, 2003.

Bugnyar, T., Heinrich, B., Ravens, *Corvus corax,* Differentiate Between Knowledgeable and Ignorant Competitors. *Proceedings of the Royal Society of London B* 272: 1641–1646, 2005.

Emery, N. J., Clayton, N., The Mentality of Crows: Convergent Evolution of Intelligence in Corvids and Apes. *Science* 306: 1903–1907, 2004.

Gallup, G. G., Chimpanzees: Self-Recognition. *Science* 167: 86–87, 1970.

Goodall, J., Tool-Using and Aimed Throwing in a Community of Free-Living Chimpanzees. *Nature* 201: 1264–166, 1964.

Griffin, D. R., *Animal Thinking.* The Harvard University Press, Cambridge, MA, 1984.

Güntürkün, O., Bugnyar, T., Cognition Without Cortex. *Trends in Cognitive Sciences* 20: 291–303, 2016.

Hare, B., Call, J., Tomasello, M., Do Chimpanzees Know What Conspecifics Know? *Animal Behaviour* 61: 139–151, 2001.

Hunt, G. R., Manufacture and Use of Hook-Tools by New Caledonian Crows. *Nature* 379: 249–251, 1996.

Izawa, K., Die Affenkultur der Rotgesichtsmakaken. In: Grzimek, B. (Ed.), *Grzimeks Enzyklopädie Säugetiere.* Kindler Verlag, München, 286–295, 1988.

Kaminski, J., Call, J., Fischer, J., Word Learning in a Domestic Dog: Evidence for 'Fast Mapping'. *Science* 304: 1682–1683, 2004.

Köhler, W., *Intelligenzprüfungen an Menschenaffen. Unveränderter Nachdruck der 2. Aufl. von 1921.* Springer, Berlin, Göttingen, Heidelberg, 1963.

Krupenye, C., Kano, F., Hirata, S., Call, J., Tomasello, M., Great Apes Anticipate That Other Individuals Will Act According to False Beliefs. *Science* 354: 110–114, 2016.

Manning, A., Dawkins, M. S., Learning and Memory. In *An Introduction to Animal Behaviour*. Sixth ed., Cambridge University Press, Cambridge, 2012.

Mendes, N., Hanus D., Call, J., Raising the Level: Orangutans Use Water as Tool. *Biology Letters* 3: 453–455, 2007.

Mercader, J., Barton, H., Gillespie, J., et al., 4300-Year-Old Chimpanzee Sites and the Origins of Percussive Stone Technology. *Proceedings of the National Academy of Sciences USA* 104: 3043–3048, 2007.

Pawlow, I. P., *Die bedingten Reflexe*. Kindler Verlag, München, 1972.

Rensch, B., Döhl, J., Wahlen zwischen zwei überschaubaren Labyrinthwegen durch einen Schimpansen. *Zeitschrift für Tierpsychologie* 25: 216–231, 1968.

Skinner, B. F., *The Behavior of Organisms*. Appleton-Century-Crofts, New York, 1938.

Van Schaik, C. P., Ancrenaz, M., Borgen, G., et al., Orangutan Cultures and the Evolution of Material Culture. *Science* 299: 102–105, 2003.

CHAPTER 6

Agrawal, A. A., Laforsch, C., Tollrian, R., Transgenerational Induction of Defences in Animals and Plants. *Nature* 401: 60–63, 1999.

Bateson, P., Gluckman, P., Hanson, M., The Biology of Developmental Plasticity and the Predictive Adaptive Response Hypothesis. *Journal of Physiology* 592: 2357–2368, 2014.

Dall, S. R. X., Houston, A. I., McNamara, J. M., The Behavioural Ecology of Personality: Consistent Individual Differences from an Adaptive Perspective. *Ecology Letters* 7: 734–739, 2004.

Dingemanse, N. J., Bouwman, K. M., van de Pol, M., et al., Variation in Personality and Behavioural Plasticity Across Four Populations of Great Tit *Parus major*. *Journal of Animal Ecology* 81: 116–126, 2012.

Freund, J., Brandmaier, A. M., Lewejohann, L., et al., Emergence of Individuality in Genetically Identical Mice. *Science* 340: 756–759, 2013.

Harlow, H. F., Harlow, M., Social Deprivation in Monkeys. *Scientific American* 207: 136–146, 1962.

Immelmann, K., Barlow, G., Petrinovitch, L., Main, M., *Behavioral Development: the Bielefeld Interdisciplinary Project*. Cambridge University Press, Cambridge, 1981.

Kaiser, S., Sachser, N., The Effects of Prenatal Stress on Behaviour: Mechanisms and Function. *Neuroscience and Biobehavioral Reviews* 29: 283–294, 2005.

Mousseau, T. A., Fox, C. W., The Adaptive Significance of Maternal Effects. *Trends in Ecology and Evolution* 13: 403–407, 1998.

Réale, D., Reader, S. M., Sol, D., McDougall, P. T., Dingemanse, N. J., Integrating Animal Temperament Within Ecology and Evolution. *Biological Reviews Cambridge Philosophical Society* 82: 291–318, 2007.

Sachser, N., Hennessy, M. B., Kaiser, S., Adaptive Modulation of Behavioural Profiles by Social Stress During Early Phases of Life and Adolescence. *Neuroscience & Biobehavioral Reviews* 35: 1518–1533, 2011.

Sachser, N., Kaiser, S., Hennessy, M. B., Behavioural Profiles Are Shaped by Social Experiences: When, How and Why. *Philosophical Transactions of the Royal Society B* 368: DOI 201203344, 2013.

Sih, A., Bell A. M., Johnson, J. C., Ziemba, R. E., Behavioral Syndromes: An Integrative Overview. *Quarterly Review of Biology* 79: 241–277, 2004.

Spear, L. P., The Adolescent Brain and Age-Related Behavioral Manifestations. *Neuroscience & Biobehavioral Reviews* 24: 417–463, 2000.

Trivers, R. L., Parent–Offspring Conflict. *American Zoologist* 14: 249–264, 1974.

Zimmermann, T. D., Kaiser, S., Hennessy, M. B., Sachser, N., Adaptive Shaping of the Behavioural and Neuroendocrine Phenotype During Adolescence. *Proceedings of the Royal Society B* 2784: DOI 101098/rspb.2016, 2017.

CHAPTER 7

Adrian, O., Brockmann, I., Hohoff, C., Sachser, N., Paternal Behaviour in Wild Guinea Pigs: A Comparative Study in Three Closely Related Species with Different Social and Mating Systems. *Journal of Zoology (London)* 265: 97–105, 2005.

Adrian, O., Sachser, N., Diversity of Social and Mating Systems in Cavies. *Journal of Mammalogy* 92: 39–53, 2011.

Alcock, J., *Animal Behavior*. Ninth ed., Sinauer, Sunderland, MA, 2009.

Bradbury, J. W., Andersson, M. B., *Sexual Selection: Testing the Alternatives*. Wiley, Chichester, 1987.

Carter, G. G., Wilkinson, G. S., Food Sharing in Vampire Bats: Reciprocal Help Predicts Donations More Than Relatedness or Harassment. *Proceedings of the Royal Society of London B* 280: DOI 20122573, 2013.

Clutton-Brock, T. H., Cooperation Between Non-Kin in Animal Societies. *Nature* 462: 51–57, 2009.

Clutton-Brock, T. H., *Mammal Societies*. John Wiley & Sons, Chichester, 2016.

Clutton-Brock, T. H., Guiness, F. E., Albon, S. D., *Red Deer. Behavior and Ecology of Two Sexes*. The University of Chicago Press, Chicago, 1982.

Darwin, C., *Die Entstehung der Arten*. Neudruck Reclam, Stuttgart, 1981 (Original 1859).

Dawkins, R., *The Selfish Gene*. Oxford University Press, Oxford, 1976.

Gilg, O., Hanski, I., Sittler, B., Cyclic Dynamics in a Simple Vertebrate Predator–Prey Community. *Science* 203: 866–868, 2003.

Hamilton, W. D., The Genetical Theory of Social Behaviour, I, II. *Journal of Theoretical Biology* 7: 1–52, 1964.

Hofer, H., East, M. L., Siblicide in Serengeti Spotted Hyenas: A Long-Term Study of Maternal Input and Cub Survival. *Behavioural Ecology and Sociobiology* 62: 341–351, 2008.

Hrdy, S. B., Infanticide Among Animals: Review, Classification, and Examination of the Implications for Reproductive Strategies of Females. *Ethology and Sociobiology* 1: 13–40, 1979.

Kappeler, P., *Verhaltensbiologie*. 3. Auflage, Springer, Heidelberg, 2012.

Keil, A., Sachser, N., Reproductive Benefits from Female Promiscuous Mating in a Small Mammal. *Ethology* 104: 897–903, 1998.

Kempenaers, B., Verheyen, G. R., Vandenbroeck, M., et al., Extra-Pair Paternity Results from Female Preference for High-Quality Males in the Blue Tit. *Nature* 357: 494–496, 1992.

Packer, C., Pusey, A. E., Infanticide in carnivores. In *Infanticide: Comparative and Evolutionary Perspectives*. Hausfater, G., Hrdy, S. B. (eds.), Aldine, New York, S. 31–42, 1984.

Sachser, N., Kaiser, S., Meerschweinchen als Sozialstrategen. *Spektrum der Wissenschaft*, 56–63, 2010.

Sherman, P. W., Nepotism and the Evolution of Alarm Calls. *Science* 197: 1246–1253, 1977.

Trivers, R. L., The Evolution of Reciprocal Altruism. *The Quarterly Review of Biology* 46: 35–57, 1971.

Trivers, R., *Social Evolution*. The Benjamin/Cummings Publishing Company, Inc., Menlo Park, CA, 1985.

Wickler, W., Seibt, U., *Das Prinzip Eigennutz. Ursachen und Konsequenzen sozialen Verhaltens*. Hoffmann und Campe, Hamburg, 1977.

Wilkinson, G. S., Reciprocal Food Sharing in the Vampire Bat. *Nature* 308: 181–184, 1984.

Wilson, E. O., *Sociobiology. The New Synthesis*. The Belknap Press of Harvard University Press, Cambridge, MA, 1975.

Wilson, M. L., Boesch, C., Fruth, B., et al., Lethal Aggression in *Pan* Is Better Explained by Adaptive Strategies Than Human Impacts. *Nature* 513: 414–417, 2014.

CHAPTER 8

De Waal, F., *Are We Smart Enough to Know How Smart Animals Are?* W. W. Norton & Company Inc., New York, 2016.

Kershenbaum, A., Bowles, A. E., Freeberg, T. M., et al., Animal Vocal Sequences: Not the Markov Chains We Thought They Were. *Proceedings of the Royal Society B* 281: DOI 10.1098/rspb.2014.1370, 2017.

Lorenz, K., *Das sogenannte Böse*. Borotha-Schoeler Verlag, Wien, 1963.

Natterson-Horowitz, B., Bowers, K., *Zoobiguity: What Animals Can Teach Us About Health and the Science of Healing*. Alfred A. Knopf, New York, 2012.

Raby, C. R., Alexis, D. M., Dickinson, A., Clayton, N. S., Planning for the Future by Western Scrub-Jays. *Nature* 445: 919–921, 2007.

Thornton, A., McAuliffe, K., Teaching in Wild Meerkats. *Science* 313: 227–229, 2006.

Tomasello, M., *Die kulturelle Entwicklung des menschlichen Denkens. Zur Evolution der Kognition*. Suhrkamp Verlag, Berlin, 2006.